AN INTRODUCTION TO
FAMILY RESOURCE MANAGEMENT

AN INTRODUCTION TO
FAMILY RESOURCE MANAGEMENT

Premavathy Seetharaman Ph.D.
Reader, Department of Resource Management,
Institute of Home Economics (University of Delhi)

Sonia Batra Ph.D.
Research Scholar, Delhi University

Preeti Mehra Ph.D.
Lecturer, Department of Resource Management,
Institute of Home Economics (University of Delhi)

CBSPD

CBS Publishers & Distributors Pvt Ltd

New Delhi • Bengaluru • Chennai • Kochi • Kolkata • Lucknow • Mumbai
Hyderabad • Jharkhand • Nagpur • Patna • Pune • Uttarakhand

An Introduction to
Family Resource Management

ISBN: 978-81-239-1186-1

First Edition: 2005
Reprint: 2015, 2019, 2023

Published by **Satish Kumar Jain** and produced by **Varun Jain** for

CBS Publishers & Distributors Pvt Ltd
4819/XI Prahlad Street, 24 Ansari Road, Daryaganj, New Delhi 110 002, India.
Ph: 011-23289259, 23266861, 23266867   Website: www.cbspd.com
Fax: 011-23243014   e-mail: delhi@cbspd.com

Corporate Office: 204 FIE, Industrial Area, Patparganj, Delhi 110 092
Ph: 011-4934 4934   Fax: 011-4934 4935

                              e-mail: publishing@cbspd.com; publicity@cbspd.com

Branches

- **Bengaluru:** Seema House 2975, 17th Cross, KR Road, Banasankari 2nd Stage, Bengaluru 560 070, Karnataka, India
  Ph: +91-80-26771678/79   Fax: +91-80-26771680   e-mail: bangalore@cbspd.com
- **Chennai:** 7, Subbaraya Street, Shenoy Nagar, Chennai 600 030, Tamil Nadu, India
  Ph: +91-44-26680620, 26681266   Fax: +91-44-42032115
- **Kochi:** 42/1325, 1326, Power House Road, Opp KSEB, Power House, Ernakulam Kochi 682 018, Kerala, India
  Ph: +91-484-4059061-65,67   Fax: +91-484-4059065   e-mail: kochi@cbspd.com
- **Kolkata:** 147, Hind Ceramics Compound, 1st Floor, Nilgunj Road, Belghoria, Kolkata-700056, West Bengal, India
  Ph: +033-25633055, 033-25633056   e-mail: kolkata@cbspd.com
- **Lucknow:** Basement, Khushnuma Complex, 7 Meerabai Marg (Behind Jawahar Bhawan),Lucknow-226001, UP, India
  Ph: +0522-4000032   e-mail: tiwari.lucknow@cbspd.com
- **Mumbai:** PWD Shed, Gala no 25/26, Ramchandra Bhatt Marg, Next to JJ Hospital Gate no. 2, Opp. Union Bank of India, Noorbaug, Mumbai-400009, Maharashtra, India
  Ph: 022-66661880/89
                              e-mail: mumbai@cbspd.com

Representatives

- Hyderabad   0-9885175004   • Jharkhand   0-9811541605   • Nagpur   0-9421945513
- Patna   0-9334159340   • Pune   0-9923910676   • Uttarakhand 0-9716462459

Printed at Sanjay Printers, Sahibabad, UP

DEPARTMENT OF HOME SCIENCE
University of Delhi

DR. (MRS.) K. KHANNA
Head
Department of Home Science

INSTITUTE OF HOME ECONOMICS
F-4, Hauz Khas Enclave,
New Delhi- 110016
Tel. : 6532402, 6510711
Fax : 6510161

Foreword

I AM EXTREMELY happy that Dr. Premavathy Seetharaman, Reader in the Department of Resource Management at Institute of Home Economics along with Dr. Sonia Batra and Dr. Preeti Mehra, research scholars, have brought out the text book "*An Introduction to Family Resource Management*", which is the culmination of their teaching and research experiences.

The book covers all aspects of Family Resource Management including Management Principles, Processes and Resources. The time, energy, money and space management has been discussed with special reference to the families and the women/home maker as the manager of these resources. The text is well illustrated with tables and figures for easy understanding.

This book specifically deals with Resource Management of families in the Indian context and therefore, will be very helpful to the students. Although it has been written as a text book for students of Home Science, the home makers/women managers also will find it very informative and helpful guide.

I am confident that this book will be an useful guide to all those women interested in becoming good managers and help them in efficiently managing their families' limited time, energy, money and space resources. I congratulate the team of authors for their commendable work of putting together their long research experience in the form of a book with meticulous details.

Dr. (Mrs.) K. Khanna

Preface ◄◄◄◄━━

FAMILY living is becoming more and more complex in its modern setting and the ever-changing family scenario. In traditional homes, home making and management activities are passed on easily from a mother to her children. However, in the present family system where the mother is occupied outside her home for longer hours, she has very little time to pass on her knowledge or train her children in the area of family resource management. Thus the present generation has become more dependent on the available literature, and this book is a humble attempt towards this goal.

For a long time, most of the teachings in this area of Home Science, books written by authors from western countries were followed. Though the system of family is more or less the same throughout the world, it would be more appropriate if the students in India are made to learn the tools and techniques of management written in Indian context. The students would be more at ease and can relate their family life to the examples given in a book if its contents are tilted towards the Indian context. Besides, Home Science education is now changing fast from its earlier focus on merely preparing the girls for having a happy married life and running a home efficiently, towards preparing girls for vocations. The present day students need a strong literature base to learn the various aspects of Home Science education which is based on facts from Indian context. There is a dearth for books of this kind. Keeping this in view, the authors have made an attempt to incorporate examples typical to Indian culture in this book on Family Resource Management.

The contents of the book include the areas of managing family resources. The fundamental concepts of management, family as a system, the resources and the motivating factors of management in a home are dealt in detail in the book. In addition to the details on these concepts, the information given in the book also deals with their inter-relationships and their relationship to management. Decisionn making, being the crux of management, is dealt with detail in a separate chapter in the book.

Some major resources like time, energy, money and space are discussed in detail in separate chapters. These resources are thoroughly analysed and presented in a format which can easily be understood and followed by the students. Simple examples from daily life are given to help the students to identify and understand these resources and their wise management in their family life. Some of the facts are also presented in the form of figures and tables to highlight them and for easy understanding.

With the changing scenario in the family life, an effort was made to discuss the role of management for the family during the various stages of its life cycle. Some major resources, their availability and the demands on them—are discussed in simple and easy language for the students to realise the significance of learning this area of the subject.

For the preparation of the manuscript, the syllabus requirements of the curriculum of Indian Universities, in particular the University of Delhi, in the area of Home Science courses for under-graduate as well as post-graduate students are kept in mind. Even though the authors have made sincere attempt to cover the relevant areas, it is possible that some of them may not have been given due significance. The authors, therefore, welcome constructive suggestions for improving the text materials in future editions.

It would not have been possible to prepare the entire material for the book without the encouragement and co-operation of our family members, in particular the support of Mr. N. Seetharaman, Mr. R. Batra and Mr. P. Mehra in every stage of this publication. We would like to record our sincere gratitude to them who helped us in all possible manner to bring this book out in a short time.

Authors

Contents ⟨⟨⟨⟵

Foreword v
Preface vii

1 INTRODUCTION TO MANAGEMENT 1-22

Philosophy of Management 1
Management is Universal 2
Management and Culture 3
The Need for Management 6
Management in Family Living 7
Characteristics of Management in the Home 8
Role of Home Management 9
What is Management 10
Management as a Discipline in Home and Business 12
Misconceptions Regarding Home Management 15
Management as an Art and Science 16
Management as a Profession 18
Women in Management 20
Ethics in Management of Family Living 21

2 MANAGERS AND MANAGEMENT 23-34

Managers and Organisations 23
Roles of a Manager 24
Roles and Responsibilities of a Homemaker 26
Decisional Roles 27

Qualities of a Homemaker 27

Characteristics of Homemaker 29

Functions of a Manager 30

Types of Managers 32

3   MANAGEMENT—PRINCIPLES AND PROCESS 35-65

History and Growth of Management 35

System's Approach to Home Management 37

Management Principles 39

The Management Process 43

Planning 44

Organising 50

Leading 54

Controlling 57

4   MOTIVATING FACTORS IN MANAGEMENT 66-87

Motivation in Management 66

Factors for Motivation 69

Values 69

Goals 79

Standards 82

Interrelatedness of Values, Goals and Standards 86

Motivating Factors and Management 86

5   DECISION MAKING—THE HEART OF MANAGEMENT 88-102

What is Decision Making 88

Steps in the Decision Making Process 92

Factors Affecting Decision Making 95

Kinds of Decisions in Families 96

How Conflicts can be Resolved 98

Complexities in Decision Making 100

Aids in Decision Making 101

6   RESOURCES IN MANAGEMENT 103-123

Meaning and Definitions of Resources 103

Role of Resources in Management 104

Characteristics of Resources 104

Classification of Resources 109

Factors Affecting the Use of Resources 116

Maximizing the Use of Resources 119

Conservation of Resources With Special Reference of Shared Resource 120

7   TIME MANAGEMENT **124-145**

Nature of Time 124

Balance in Time Use 125

Tools in Time Management 130

Factors to Consider in Making Time and Activity Plans 135

Everyday Jobs and Activities 136

Weekly and Special Tasks and Activities 136

Seasonal Tasks and Activities 137

Sample Time Schedules 138

Time Demands During Different Stages of the Family Life Cycle 142

Control of Time Plans 143

Evaluating Time Plans 144

Tips for Managing Time 144

8   ENERGY MANAGEMENT **146-172**

Relation of Energy to the Stages of the Family Life Cycle 146

Classification of Efforts Used in Homemaking Activities 147

Posture in Housework 154

Fatigue; Types, Causes and Remedies 155

How to Avoid Fatigue 159

Remedies for Frustration Fatigue 160

Techniques of Work Simplification 161

Management Process as Applied to Energy Resource 169

9   MONEY MANAGEMENT **173-194**

Concept of Income 173

Sources and Types of Family Income 174

Methods of Supplementing Family Income 177

Steps in Money Management 178

Budgeting 179

Steps in Making a Budget 180

Controlling the Use of Income 184

Types of Records 187

Evaluation 187

What is Savings 188

Advantages of Savings 189

Savings and Investments 190

Guidelines of Sound Investment 190

Avenues for Savings and Investments 191

10 SPACE MANAGEMENT **195-220**

Selecting a House 195

What Kind of a House Does the Family Need? 196

Factors that Affect the Choice of a House 196

Spatial Planning 197

Controlling the Use of Space in a Home 201

Principles of Good Storage 205

Sanitation 209

Aesthetic Arrangement of Space 209

Accessories 210

Flower Arrangement 211

Rangoli: The Art of Floor Decoration 215

Evaluating the Use of Space 219

11 FAMILY LIFE CYCLE **221-241**

The Environments Surrounding the Family 222

The Concept of the Family 223

Role of the Family 224

Concept of Family Life Cycle 227

Importance of Recognizing the Family Life Cycle 228

Analysis of Family Life Cycle 229

Recent Trends in Family Life Cycle 230

Impact of Recent Changes in the Family System on the Households 231

Small Family vs Large Family 232

Recent Changes in Family Life 232

Management During Family Life Cycle 233

Handling Demands During the Various Stages of Family Life Cycle 236

Importance of Understanding Family Life Cycle 241

Bibliography 243-246

Index 247

1

Introduction to Management <<<

MANAGEMENT is a purposeful activity, undertaken to achieve set targets. Everyone of our life is full of goals and when there is a goal, management is the right action to ensure its accomplishment. Management is an activity-oriented process in which results are reached systematically by combining the efforts of people. Thus management is an art of getting things done by self or by others. A manager or a person, who undertakes the management activity, integrates the activities and unifies efforts, resulting in the implementation of the planned activities for the achievement of goals.

Management is pervasive to every aspect of human living and is found universal in nature. We can say that management is an activity-based process, which is also integrative in nature. Though management is a separate discipline, today it has drawn knowledge from other disciplines and has emerged as an inter-disciplinary field of study. In this chapter, we will focus upon these characteristics of management and will make an attempt to define it. In this book the focus of management is in relation to home, therefore, we will also explore the characteristics, role, definition of home management and its distinction from the other forms of management. Also we will discuss some basic concepts and issues like philosophy, universality, relation to science and art, professionalism and ethics in management.

PHILOSOPHY OF MANAGEMENT

Management is a combination of fundamental philosophical thinking with rather simple techniques. To get success in this field, it is important not only to look for the techniques, but also to understand its depth and challenges. To illustrate this, let us take the matter of managing family money. Budgeting techniques are quite simple and commonly used by everyone. On the other hand, there is a lack of awareness about the philosophical background for the financial choices made during the preparation of the budget. Lack of this awareness reduces the chances of success of the budget and makes it impractical and questionable. The way a family uses its money, whether blindly or consciously, is an expression of its philosophy of living. We can realise this in our day-to-day life. For example, families giving more emphasis on up-to-dateness would spend more at present than by saving to ensure a secure future. On the contrary, people who want to have a secure future would save more money than to spend on their daily needs. They may even sacrifice some of their present needs to ensure a secure future. They might utilise more of the other resources like time and energy, instead of money, to meet their needs. Therefore, it is necessary for the family to formulate its philosophy clearly so that it becomes easy for the family members to channelise their major resources, like time, energy and money, for productive purposes. Another philosophical concept which underlines family's use of money, concerns the rights of all members of the family to economic resources, in relation to the other resources. This may also get reflected in the way the family allots the money resource in its budget.

Thus, it is clear that in order to get success in the utilisation of resources, it is necessary that fundamental values must be weighed carefully. Once the philosophical choices are made and their underlying values are accepted, the technique of management falls in line readily. Therefore, management is the key to the leadership to decide on the direction in the use of its resources for the smooth running of a home. In this effort, the underlying values and standards provide the necessary motivation for the achievement of goals.

MANAGEMENT IS UNIVERSAL

As a human being, every one of us is a member of an organisation or an enterprise, be it a home, sports club, college, religious association or a business. The various organisations or groups differ in one or more respect. However, in all these organisations, among the various areas of human activities, the act of managing certainly finds an important place. This is so, because all the members of these organisations, even while pursuing personal interests, work together to achieve common objectives of that organisation. Thus we can say that the membership of these organisation is characterized by mutual cooperation and pursuit of some common objectives. Thus, Management, which in nut shell, is a process of achievement of goals is universal to all organisations.

Management, as can be seen in Figure 1.1, is an essential part in all organised activities and human endeavors. It is a basic task of any organisation where two or more people pursue common objectives with mutual cooperation. Management is, thus, as an important component in a highly specialised field like business as in the last realised (or the most basic) institution like a household. While management is a conscious effort in a business field, it is a spontaneous act on the part of a homemaker. Therefore, we can say that management is an essential component of every field of living, ranging from the most fundamental unit of the society to the most complex organisation, as observed in our modern lives.

Just like management is universal to all organisations, it is pervasive to every family and its every aspect universally. Every family manages its resources to achieve its goals. However, some families manage their resources better than others and thus, achieve more goals by spending lesser resources. This signifies their understanding of management knowledge and ability to practice it better than those families which cannot derive much satisfaction from their resources. However, among all better managed families around the world, the process of management differs due to the influence of culture on management. Never the less, management is universal to all families in every corner of the world. Thus, there is a great need to understand and apply, the science and art of management.

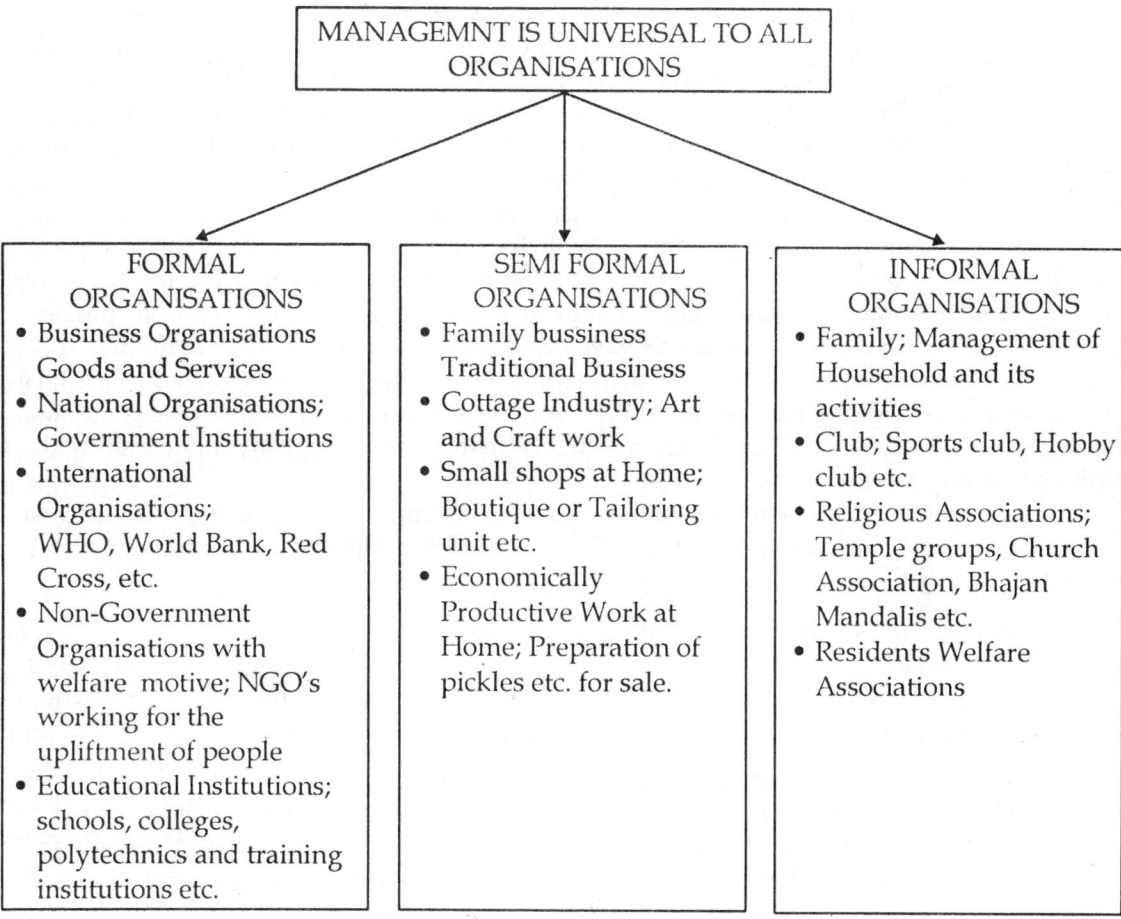

Fig. 1.1: Management is Universal to all Organisations

MANAGEMENT AND CULTURE

Management and culture influence each other. Similar to management, culture is also found to be universal. Both have a social function. Both are socially accountable and closely embedded. Payne[1] identifies the major points about the concept of culture as, *it is only one way of looking at the process of organizing; The central core of concept includes ideology, beliefs and values; These ideologies, values and beliefs, and the patterns of behaviour that flow from them are created and perpetuated through means embedded in social structures and systems.*

Thus, culture in the organisational context can be defined as the sum of a group or people's way of thinking, believing, feeling and acting. The understanding of culture provides better alternatives for decision making by managers. However, it also puts restrictions on the available alternatives and their managers need to conform to the cultural values of the organisation. This is so because of two reasons. First, no organisation can be isolated from its cultural environment, that is, organisation as a social unit must operate within the framework of the larger cultural

[1]Payne, R. (1992). Some thoughts on culture and the role of the personnel function. Academy of Management Conference. Bradford, Yorkshire.

system. A congruency has to be maintained with the total culture. Secondly, organisation may be considered as a subculture within the framework of the broader culture. Every organisation thus develops its norms and cultural patterns of behaviour within the context of the larger cultural environment. Thus both the organisational culture and the environmental culture including the national culture must synchronize with one another. Therefore management under each cultural situation differs and is, therefore, said to have a cultural bias.

Cultural characteristic is typical to every organization as it expresses its individuality and uniqueness. The values, orientation, attitudes and beliefs of individuals, which people bring with them to the job, also influence the culture of an organisation. They provide the basis for judgement, analysis and discrimination. This in turn helps in making choices from possible behavioural alternatives. Also individuals have preconceived notions of what **ought** and what **ought not** to be done. These notions contain interpretations of right and wrong. They imply that certain behaviour is preferred over the others. As a result, culture can cloud objectivity and rationality. Thus decision-making which is a crux of management is greatly influenced by the culture of the people, organisation and even, nation.

Hofstede[2] confirmed that national culture has a major impact on managers' work related values and attitudes. He found that managers vary on the following four dimensions of national cultures:

- Individualism versus Collectivism
- Power Distance
- Uncertainty Avoidance
- Masculinity versus Femininity

The way the managers manage their organisation depends on how individualistic or collective they are or how much power distance they adhere to; or how much they try to avoid uncertainty rather than take risks; or how feminine (caring and emotional) or masculine (assertive) they are. Table 1.1 shows how these cultural dimensions influence the management both at work place and in the family.

Every culture also has its own set of values, which influences its members. Many managers often find a conflict between their own values and those of others because of the difference in their cultural backgrounds. Management in the workplace, as seen by managers around the world varies considerably. Alkhafaji and Abbass[3] found that in Australia, managers place a high value on honour, loyalty, tolerance and compassion indicating a strong human approach. This results in stronger ties between business partners but limits expansion and innovation of new products worldwide. Japan, on the other hand places more value on the work, and therefore, is always on the leading edge as far as their products are concerned. The Japanese are trained to work hard to achieve their goals and thus, hard work has contributed to their company to become the best in the world.

In the case of America, a higher value on the individual rather than on the organisation is placed by the people. Creativity flourishes here through ambition, which is passed from one generation to the other. They are more concerned with individual ability, skill, ambition, capability and drive

---

[2]Hofstede, G. (1991). *Cultures and Organisation: Software of the Mind*. U.K.; McGraw Hill.
[3]Alkhafaji and Abbass, F. (1992). *Competitive Global Management: Principles and Strategies*. New Delhi; Vanity Book International.

in getting the job done. This leads to a competition within the workplace rather than with the other organisations. Table 1.2 shows how mangers from India, USA and Japan hold significantly different values due to the differences in their national cultures.

TABLE 1.1: CULTURE DIMENSIONS AND MANAGEMENT IN THE FAMILY AND WORKPLACE*

Culture Dimension	Management in the Family	Management at Work Place
Power Distance (PD)		
• Small PD societies	-Children encouraged to have a will of their own -Parents treated as equals	-Hierarchy means an inequality of roles -Ideal boss is a resourceful democrat
• Large PD societies	-Children educated towards obedience -Parents treated as superiors	-Hierarchy means existential inequality -Ideal boss is benevolent autocrat
Collectivism Vs Individualism		
• Collective societies	-Education towards "we" consciousness -Obligation to family -Harmony, respect and shame	-Value standards differ for in-group and out-group -Relationship prevails over task -Other people are seen as members of their group
• Individualistic societies	-Education towards "I" consciousness -Obligation to self -Self- interest, guilt, self actualization	-Same value standards apply to all: universalism -Task prevails over relationship -Other people seen as potential resources
Uncertainty avoidance (UA)		
• Weak UA societies	-Ease, indolence, low stress -Aggression and emotions not shown	-Dislike of rules -Less formalization and standardization
• Strong UA societies	-Higher anxiety and stress -Show of aggression and emotion acceptable	-Emotional need for rules -More formalization and standardization
Femininity vs masculinity		
• Feminine societies	-Stress on relationship -Solidarity -Resolution of conflicts by compromise and negotiation	-Assertiveness ridiculed -Undersell your self -Stress on life quality -Intuition
• Masculine societies	-Stress on achievement -Competition -Resolution of conflicts by fighting them out	-Assertiveness appreciated -Oversell yourself -Stress on careers -Decisiveness

*Source: Hofstede, G. (1991). *Culture and Organisations: Software of the Mind*. Berkshire, U.K; McGraw Hill.

These cultural differences also influence the family management practices globally. Though all families utilise the management knowledge, their decision-making, interpersonal relationships, personal preferences, even goals and morals are governed by their cultural value. For example, children in American families are treated as equal to parents and are encouraged to have a will of their own, whereas, in India, they are educated towards obedience and respect for all elders in the family. Japanese families educate their children towards "we" consciousness and inculcate in them an obligation to family. They also stress on relationships and solidarity and resolve their conflicts by compromise and negotiation. On the other hand, families in America, stress on achievement and competition and resolve their conflicts by fighting them out.

Thus, in management culture plays an important role and can be traced to the responses to common human problems. The culture has a strong influence on the way people behave and manage their organisations even in cases of their own home. Therefore, management is not only universal but, is also culture biased.

TABLE 1.2: COMPARATIVE CULTURAL VALUE PROFILES OF INDIAN MANAGERS WITH THOSE OF U.S.A. AND JAPAN*

S.No.	Indian Cultural values	USA Cultural values	Japanese Cultural values
1.	Life is preplanned, human action is predetermined.	Individuals can be influenced.	Groupism.
2.	I need to adjust, human action is pre- determined.	I can change work to achieve objective and commitment to organisation.	Homogeneity.
3.	Decisions flow from the experience and wisdom of authorities.	Data - based decisions and they are healthy.	Confusion ethic
4.	Deference to age and seniority, suppression of negative feelings.	I can disagree without being disagreeable.	High educational levels.
5.	Joint family and authoritarian values.	Protestant ethics.	
6	Self-realization.	Authentic collaboration.	

*Source: Chakroborty, S. K. (1993). *Management by Values: Towards Cultural Congruences*. New Delhi; Oxford University Press.

THE NEED FOR MANAGEMENT

Though management is as old as human society, until the modern era, little was done to establish it as a body of knowledge. For several thousand years of actual management, practice proceeded management thought. The need for management arose when the resources involved became scarce and limited, both quantitatively and qualitatively. This has led the man to go around, and find ways to increase their availability. As the environment became more and more complex and with increasing competition from his fellow men, he had to look for many possibilities in the use of his resources. Thus, the focus became choosing the right course of action, whereby goal is achieved using the least amount of resources. The industrial revolution compelled businessmen to address management as an independent subject for the first time. Though, the complex organisations emerged in early 1900s, it was only after the World War II, management emerged as a field of study.

Management in its simplest form is said to be the process of directing operations of an organisation, to realise the established goals. Management is the development of people. It is the management, which provides a coordinating link to the people engaged in the achievement of common objectives.

The concept of management involves planned use of resources directed towards the achievement of desired ends. This involves the weighing of values and the making of series of decisions. As we have seen earlier, culture influences the management decision. Within the same culture also, values may vary. For example, one family may be giving more value to exhibit a higher standard of living. While doing so, they might spend more on luxurious items than on the essential items or on providing higher or specialised education. On the other hand, family valuing education may be spending more on their children's educational materials and have a simple way of living. Thus the values of families having similar cultural backgrounds, may differ from one another. However, both the families may act in an orderly manner in which the resources and goals are dealt with, and this becomes the purpose of their management.

Thus, management is important for the successful accomplishment of goals of the organisation where it is practiced. It also acts as motivator to achieve goals by focussing all activities and efforts towards them.

MANAGEMENT IN FAMILY LIVING

Management in the context on the family is the natural outgrowth of human associations and interactions. Its ultimate aim is to provide for optimal development of its individual members. Management of family allows us to overcome our individual limitations. Through the combination of individual efforts and resources, we achieve far more than what we could do independently. For example, if a woman cooks meals with the help of her children and helps them in their studies, both tasks are accomplished faster while leading to a greater satisfaction and happiness to all those involved. This synergy indicates that the results of combined efforts are greater than the individual efforts which is an important outcome of management. It also develops a well-coordinated system of work.

In management, it is the task of the manager to select, design and maintain a suitable environment in which all the individuals as members of its group are able to accomplish their objectives and goals. In a home situation, effective management to a large extent depends on the managerial ability, interest and leadership quality of the homemaker, so as to motivate the family members to act in the right direction for achieving the desired goals. For instance, in every home, the homemaker, who acts as a manager should know how to get all the jobs done and how to bring up happy and healthy children. She should create a congenial atmosphere in the house so that she can provide comfort, convenience, good health and happiness to all the family members. Therefore, we can say that this is the ultimate goal towards which all the activities of management in the home are directed.

By working together, the home-manager is able to achieve the fruits for all the members of the family when the managerial activities are carried out. These activities would involve many functions like planning, organising, controlling, leading, delegating, supervising and so on. While doing so, the homemaker also prepares a set of guidelines for using the resources available to her and her family members in the best possible way.

However, the final objective of all these activities and managerial functions is the same, that is, the attainment of family goals. Thus, goals also act as and become the endpoints of all managerial activities. Figure 1.2 indicates that management is a process involving activities, through which action is initiated and resources are used for achieving a goal. For this purpose, certain guidelines can be formulated by every manager. Also, for the 'sure' attainment of desired goals it becomes essential to 'plan', 'organise', 'coordinate' and 'control' all the activities, so that the resources are not wasted. This organised activity can otherwise be called as the act of managing. In the home it calls for 'integration' which is achieved by organising the activities into larger network of action to achieve the major objective of family development.

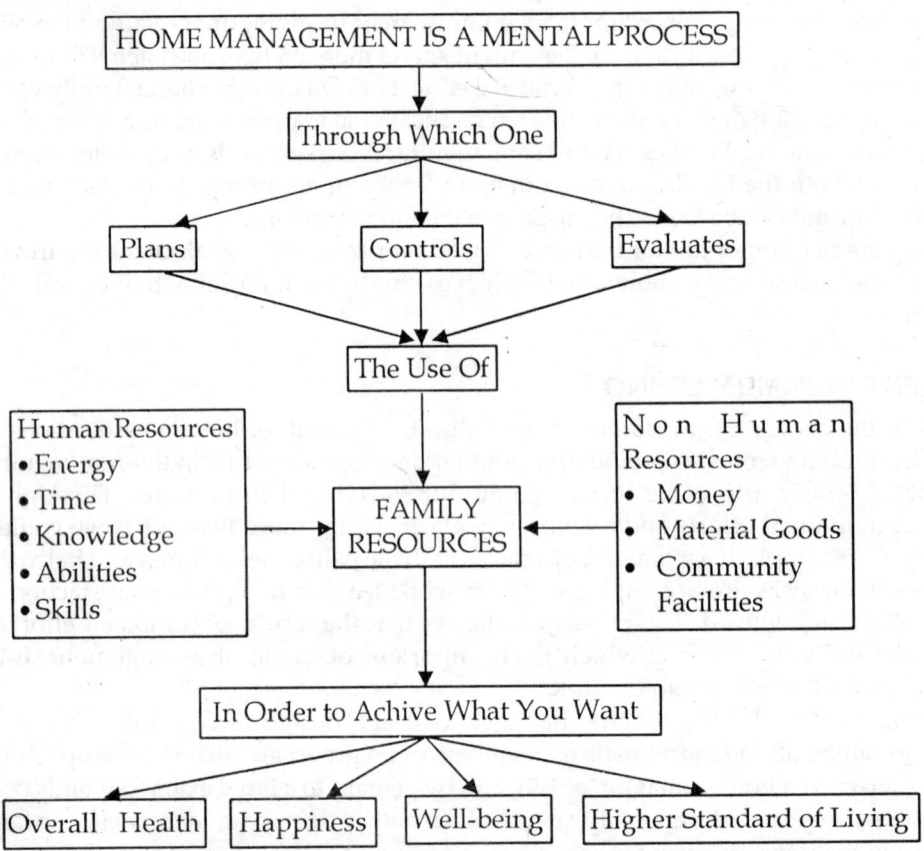

Fig. 1.2: Management Process as Applied in the Home

CHARACTERISTICS OF MANAGEMENT IN THE HOME

The primary characteristic of management is that it is universal to every home, i.e. it takes place in every home. Every family has to manage its resources to achieve its goals and work towards the satisfaction of all the members. Bigger families need to manage as they have a large amount of money and more number of people to satisfy, but even smaller families with very moderate incomes also need to manage their resources to get the best out of their means.

Though management takes place in every home, its quality varies from one family to another. The quality of management in the home is dependent on the intelligence, abilities and skills of its manager. Therefore, the study of home management tries to empower home manager with those skills, abilities and knowledge that will help them in improving the quality of management, irrespective of the quantity of available resources at their disposal.

Home management is pervasive to every aspect of family living, i.e. it influences and is influenced by every aspect concerning family as a whole as well as its individual members. Starting from the expenditure of the material resource like money to fulfill the happiness, desires and aspirations of all the members, it continue to play its role in the expenditure of the other non-material resources like knowledge, skill, ability, energy etc. All these come within the purview of home management and all of them simultaneously influence setting up of the family goals and the expenditure of their resource in their achievements.

The responsibility of transfer of home management skills to the younger members of the family is another important characteristic. Children learn to manage the family affairs at home through observation and practice, and pick up these skills of management in the process. They will apply these skills at a later stage to bigger and more formal organisations, with which they are likely to come in contact. We can thus find that children who learn to organise their things properly in their rooms at home, apply their space management skills in their school or college and later at their work place as well.

The process for democratic decision-making is another feature of home management. At home, home-managers take decisions in collaboration with the family members. After considering the opinion of every family member, the final decision is made. Thus, democracy is an essential ingredient in home management. It also highlights the importance of group decisions over individual decisions and how individual decisions could be converted into the group decisions. This can be observed in many modern family situations. In the case of a teenager girl who wants to pursue higher technical studies in a specific area, will have to convince not only her mother or father alone but will have to justifiably communicate and indicate her growing monetary demands on this account to all other members of the family. Only then her desire to pursue technical course will not be criticized or objected to, but will be willingly accepted by the family.

Home management is a process of achieving family goals by utilising family resources. In doing so, it adheres to the science and art of management. If management is a process of achievement of goals with the help of available resources, in home management, homemaker and her family members set their family goals. They all work together for the achievement of these family goals with the help of the resources available at their disposal. Thus, the family resources will be utilised for attaining family goals, as in the case of any other managerial situation.

ROLE OF HOME MANAGEMENT

The characteristics of home management, as seen earlier, enable it in playing a very important role in family living. Now, let us highlight some of the most important roles of home management.

- Home management ensures effective and optimal utilisation of family resources by eliminating the wastage of any important resource. It directs the use of family resources towards the accomplishment of family goals. In this way, it makes a conscious effort to increase the achievement of goals. When a conscious effort is made through managerial process, the family prepares a list of all possible and necessary goals. When such a process is not adopted, it is possible that some of the family goals are left out.

None of the goals of the family goes unnoticed when it adopts the right managerial process. Thus, more goals are attained by spending fewer resources through management.

- Home management makes possible the systematic accomplishment of work. All the efforts are unified and directed towards the achievement of work goals. The systematic way of working ensures that all members of the family participate in carrying out the activities. Unified effort of the family members also creates a harmonious relationship among them. It leads to a situation where the human resources of the family members are utilised in a constructive and fruitful manner, thereby leading to the accomplishment of more number of goals.

- When all the members of the family are involved in the management of the home, it establishes satisfaction among them. Satisfaction is the key and the ultimate goal of home management which is common to all families. This is true in the case of families where resources are highly limited with more demands. Through home management, family members realise the constraint and problems faced during the utilization of resources, and therefore, exhibit satisfaction, even when some of their needs are not met.

- Home management helps the homemaker in the effective utilization of family resources such as money, time and energy of all the family members in home situations. For example, children can be encouraged to spare some time from their schedule to help their mothers in carrying out the household tasks. She in turn, can use some of her time in sharing the activities of her children. This will lead to better management and will encourage cooperation and harmony at home, thereby maximising satisfaction and minimising the utilisation of scarce resources. In the process, children also learn to manage and exploit their resources well. This imparts training to them, which they can utilize successfully in their later part of life as homemakers themselves.

- Home management along with the achievement of family goals, helps all its members in reaching their personal and professional goals as well. It ensures a secure home base for all its members, irrespective of the quantitative or qualitative limitations of the available resources. As discussed earlier, satisfaction attained through management, leads to an overall personality development of the family members. It helps them to shine in their future endeavours.

- Home management encourages creative skills and abilities of family members. These creative skills can become a profession or a life long pursuit for some of its members. Creativity also gives extra footage to everyone in her/his professional life and its importance can be realised and reinforced at home while managing home in the initial stages. This can be observed among the accomplished workers who hail from well-managed homes, or from those who manage their family business with higher success rates.

WHAT IS MANAGEMENT

There are many definitions of management, some are simple, others a little elaborate. Eminent professional management writers and thinkers have provided varied definitions for management. Since the major purpose of this book is home management, special emphasis is made to relate these to the management of home.

Gross and Crandall[4] point out in the simplest way that management is *using what you have to get what you want*. In this reference to management of home, *what you have* consists of the available resources of the family members. These include not only time, energy, money and material goods, but also knowledge, interests, abilities, skills and attitudes of the family members and the available community facilities. Management can therefore be defined as *the mental process of utilizing the available resources to achieve what you want in life*. It challenges people to use their resources for the purpose they consider important. According to them *home management to a great extent than many other fields of subject matter, cuts across the area of living, for it is concerned with the ways in which a family uses all its resources*. This definition emphasises the management process as using the resources to attain the family goals.

Gross and Crandall further elaborate that *management consists of a series of decision making process of using resources to achieve goals. This process consists of three more or less consecutive steps such as planning, organising and controlling the various elements of the plan while carrying it through, whether it is executed by oneself or by others and evaluating the results preparatory to future planning*. This definition underline 'mental process' and 'sequence of steps' as important parts of management as a process.

A similar definition is given by Nickell and Dorsey[5] who points out that in general terms the concept of *management may be said to be a planned activity towards the accomplishment of desired ends. It involves the weighing of values and the making of decisions*. In this definition again emphasis is laid on the role of decision making and management being a mental process.

Management has been called *the art of getting things done through people* by Mary Parker Follet. She called attention to the fact that 'manager achieves organisational goals by arranging for others to perform whatever task may be necessary, rather than doing everything himself.' This definition holds good more for business organisations than for the household since home management is much more than this. So much more, in fact that such a definition may not be able to satisfy all the family members. In a home, the manager, i.e. the homemaker, participates equally in performing all the household activities.

L. C. Hart has a different view; management means control and control means action, therefore, management succeeds not by what it has accomplished in the part but by its ability to control what is happening at present and what is going to happen in the future. The heart of management is "change". So its main purpose is to achieve goals and in doing so it brings about changes. However, the steps in the process of management remain the same, it is the goals and the kind of resources and the demands upon these resources may vary. This definition can hold true in the case of the house too.

Brech gives a different thought to it, and defines management *as a social process entailing responsibility for the effective planning and regulation of the operation of the enterprise*. Such a responsibility involves two things. i.e.

- Judgement and decision in determining plans and using data to control performance and progress against plans, and
- The guidance, integration, inspiration and supervision of the personnel comprising the enterprise and carrying out its operation.

---

[4]Gross, I. H. and Crandall, E. W. (1954). *Management for Modern Families*. Delhi; Sterling Publishers Pvt. Ltd.
[5]Nickell, P. and Dorsey, J. M. (1991). *Management in Family Living*. New Delhi; Winley Eastern Ltd.

Even though this definition relates, directly to management in business, it holds well in a home situation too. In the management of home also, it calls for a social process with effective planning, regulation, judgement, decision making, guidance, integration, supervision and so on.

According to Henry Fayol, the main purpose of the management is *to forecast and to plan, to organise, to command, to coordinate and to control, to foresee and provide means for examining the future and drawing up the plan of action.* In other words, to organise means building up the dual structure, material and human resources of an undertaking; to command means combining together, unifying and harmonising all activities and efforts; and to control means seeing that everything occurs in conformity with established rules and expressed commands. This, to some extent is true in the case of home management too.

Urwick and Brech have rightly observed that *no ideology, no ism, and no political theory can win a greater output with less effort from a given complex of human and material resources as only sound management, and it is on such greater output that a higher standard of life, more leisure, more amenities for all must necessary be founded.* Isn't this true, when a homemaker makes efforts to manage her home?

In the words of Chatterjee, *management is an activity based process, composed of some basic functions for getting the objectives of any enterprise accomplished with and through the efforts of its personnel.* In the case of the house, the personnel refers to its members.

In the words of Gudjansson, *household management is in all countries the most common occupation employing most people handling the most money and is of fundamental importance for the health of the people.*

In home management, a home in which goals (ends) are being attained with some degree of satisfaction may be considered as a well managed home where management is practiced in an orderly manner. Home management is the vital factor in every family contributing to the overall health, happiness and well being and higher standard of living for the family members. In simpler terms, home management is defined as the mental process of utilising the available resources to achieve what you want in life.

Home management, according to Kotzin, is a practical science. It cannot do away with a concept of degree or measurement. It is a practical science such as medicine, where measurement of normal or abnormal health show a state of well-being in terms of degree.

The changes as we experience in our day-to-day life calls for a change in the way we manage our home and its resources, that are available for our use. The fundamental purpose of management is to bring about change in an orderly way. Managers are to consider what might be done better in future than what was done in the past and predict what changes or no change in procedure are likely to be required in the future. Thus one of the important thing concerning management is that it is a means of adjusting to change. These changes occur both within the family as well as in the external conditions.

MANAGEMENT AS A DISCIPLINE IN HOME AND BUSINESS

Management is a skill in running any organisational system. The organisation can be as simple as a family or as complex as a business center. Unfortunately, all the references to management are restricted specifically to the organisations in the private sector industry. This is probably because the greatest noise and initiative related to the development of management skills and the motive for achieving greater success and profits initially came from them. Though, management is used in running every organisation, there are certain differences in their management depending upon their organisational type (formal, semi-formal or informal), motive (profit or development or satisfaction), size (number of people involved or quantity of resources involved) and complexity (size or composition). Home and business are two ends of a spectrum varying upon the different aspects of organisational structure, where management plays a significant role in the achievement of their goals.

- First of all, just as business is a system created to satisfy needs and desires of a society, home is a system created to satisfy needs and desires of its family members.
- Management process is applied in both business management and in home management. both utilize the processes like planning, organising, leading, controlling and evaluating.
- The main purpose of both business and home management is to bring about an orderly change by utilizing the resources at its disposal. The kind of resources may vary but the main purpose remains the same.
- Management functions, skills and techniques are applied in both home and business situations. They are directed towards achievement of certain objectives or goals and are culture biased to some extent.
- Last but not the least, the principle of division of labour, authority and responsibility as the segments of management, are applied both in the home and business affairs and activities.

Thus, we see that there are a number of similarities between home and business management. At the same time there are also many differences among these two important organisations of the society.

- Home management involves simple planning, organising, controlling and evaluation of resources available to reach family goals. It is the side of family living. It is the force that puts the machinery of home making into action and keep it going. On the other hand, business management may be defined as an assemblage or combination of things or parts forming a complex whole, assemblage or set of correlated body of methods or a complex plan or procedure.
- While home management is limited to small-scale production for self-consumption and revolves around the home and its members only, business management is carried out at a larger scale, as many people are involved in it. It includes large-scale production and marketing etc. In business management, the number of people can increase or decrease according to the requirements. People are hired when needed and can be fired out when they are no more useful. In home management the members of the family cannot be removed as per convenience. The relationship between family members is very close and long lasting, whereas in business, bond between workers are not very strong and is only for the duration they are working together. Also, business management focuses mainly on the material resources like machinery, equipment etc., whereas, in home management emphasis is more on human resources.
- In home management the purpose is not to make profits but to develop human resources *i.e.*, it totally aims at the personality development of its members, whereas, business management is undertaken with the view to make profits. Basic human needs are neglected in business management while they are given top priority in home management. In business management machinery is used and its contribution is highly emphasized whereas in home management contribution of its members is given top priority. Even if machinery is used in the home, it is done in the interests of the family members and their productivity.

- In home management family activities take place from time to time or may overlap each other, as per the needs of the family members. Whereas, in business management, definite steps are followed regularity and continuously in all its dealings with clients as well as in the activities like production, sales etc. The activities of business management are geared always towards profits, whereas in home management, these vary according to the stages of family life cycle. In the beginning stage the efforts are made to set up a family and then on child-bearing, then activities centre around upbringing and educating the children, and finally the family goals are focussed on the marriage and vocational settlement of the children. Thus, the family activities keep varying, whereas in business, it may not be so.

- In home management there is no risk and the focus is on the human development of its members alone. The family can easily make efforts to provide satisfaction to its members without taking any risks, with little influence exerted by the environmental or external forces. However, business management carries the element of uncertainty and risk. Risk in terms of investments made in business and is dependent on the external environment forces, many of which are not in the control of the manager.

- In home management goals are unlimited, diffused and abstract in nature. The focus is more on satisfaction and all round development of the family members rather than on profits. In business management, goals are clear and limited, all targets set and objectives are very specific. The major goal is earning more and more profits and all the activities are centered around it.

- In home management untrained family members have to be utilized and trained but business management involves specialized people, trained in their field. No specialisation of any kind is desired or expected in a home situation. Also, duties in business management are fixed and not inter-changeable, but at home they are more flexible and interchangeable and the emphasis is laid upon the capacities and skills of the family members.

- At home, family members contribute in running the household for the welfare of the family and as such no fixed amount of money is paid for their contributions. Family members, instead, enjoy and receive services and other material goods like food, clothing, etc. and not monetary gains. In business management the workers are paid fixed wages periodically i.e. daily, weekly or monthly.

- Home management aims at social health of all its members and satisfaction of their needs and wants, therefore, emphasis is laid on intrinsic values such as love and affection. On the other hand, business management is profit oriented, it involves instrumental values like discipline for profit.

- At home the standards are flexible and qualitative in nature while standards maintained by business management are rigid and quantitative in nature. Moreover, in business, decision making is solely the responsibility of top management. At home however, democratic decisions are generally made. It is seen in the house that decisions are mostly taken by the elders of the family in consultation with the other members of the family.

- At home, instead of directing, guiding is undertaken and the evaluation is always not on the work output, but also on the development of the person doing the work.

Evaluation thus, is much more flexible at home because people are more important than the work output. In business managers give directions to their subordinates on matters related to work. Evaluation of work is done on the basis of its quantity or quality of work output, which is measured against the given work. At business, formal control system and a work schedule is followed accompanied by regular appraisals while no formal control is exercised at home.

- At home division of work is done keeping in mind the age, sex, interest and the abilities of the family members while in business it is according to the specialisation of the person. Also, at home the same work can be done by different people at different times while at business everyone's role is fixed and has to be performed by only that person.

MISCONCEPTIONS REGARDING HOME MANAGEMENT

Although management and its importance are increasingly understood, there are still many misconceptions especially regarding home management as to what it is, thereby creating obstacles in the way of its realisation. Proper and thorough analysis of the misconceptions are likely to throw light on the true value of management in the home.

♦ *Misconception*: Home management is mere performance of work.

Truth: Home management involves decision-making and is a mental process behind all activities.

Management in the home is a part of the fabric of living. Its threads are interwoven because decisions for the use of resources are made, whether the family is at work or at play. It is easy to see how management and performance of work have become so confused, that for a decision does not become apparent until it is translated into action—which in the home frequently means physical activity. Thus, specific situations and activities must be used to illustrate managerial decisions, but an activity itself is managerial only when it requires fresh decisions. So we can say that management is not mere performance of work, and in a home, the manager and the performer are the same person. It is especially difficult to delink and recognise the managerial aspect of carrying out a job in a home situation by the homemaker. This can be illustrated by a simple example-every day, the homemaker decides the menu and prepares the food accordingly. All related decisions like the purchase of groceries, vegetables, milk and the other ingredients are planned and bought mostly by the homemaker herself. These decisions generally remain unnoticed, and therefore, not recognised.

♦ *Misconception*: Home management is limited to the leader of a group i.e. the homemaker.

Truth: Home management is not limited to the homemaker alone, as other members of the family are also involved in the decision making, as well as in carrying out the household activities.

Another misconception concerning management in the home is that, in each family there should be one person who is the 'manager'. This leads to the unpleasant feeling, leading to the misconception that this individual has "too much power" and "manage" the other members of the group. It is true that in every managerial situation, there is usually a leader, specially the woman in a home situation, who is old and experienced enough to make decisions. However, the other family members do help the homemaker while making decisions on the choice of family resources and group goals. Besides, every individual member in a family is free to make choices regarding their individual resources like time, energy etc. The leadership role in a family may also shift from time to time. The role of a leader to household activities slips from one member of the family to another, as the situation demands. This can also be seen in the situation mentioned earlier, that at times, the shopping may be done by the husband or the children, and therefore, may take the leadership role in deciding the menu.

♦ *Misconception*: Good home managers are born not trained.

Truth: Good home managers are trained.

The other misconception about home management is that good managers are "born"and not made or "trained". However, there is a positive correlation between the amount and the type of education and the homemaker's score on a managerial yardstick. In addition, those homemakers who, when trained and educated in home management, act as natural leaders. Much learning of this nature can also come through the day-to-day contacts in a family where the need for effective management is an intelligent concern. Generally, it is said that managerial abilities at home are 'caught' as well as 'taught'. This can be observed when an experienced homemaker performs and manages better as compared to a young and recently married homemaker.

♦ *Misconception*: Management is an end in itself.

Truth: Management is a means to achieve goals, and not an end in itself.

Still another misconception concerning home management has come into focus because some families have made a fetish of management itself. These families are unable to see that management is not an end in itself but is a means to achieving family goals. Some even feel that if they try to improve their management practices, they will become too mechanical and that all pleasure will become be taken out of life. Here again one need to restudy the definition of home management to see that the fundamental purpose of management is to achieve family goals, therefore, management is a means to achieve the desired ends, i.e. the goals.

♦ *Misconception*: Family goals are dictated by management.

Truth: Family goals are governed by the family values and standards. Management selects ways and means of reaching them.

The last misconception to be mentioned here is that home management dictates family's goals, whereas, it is the values and standards held by the families which do so. This however, emphasizes that careful selection of goals is as vital as, or perhaps more vital than the careful selection of ways to reach them. Becoming aware of the implications of what one has chosen is important.

MANAGEMENT AS AN ART AND SCIENCE

It has already been said that managing involves 'organisation' as one of its most important functions. While managing any organisation, it calls for skill to carry out the activities in a systematic way. Therefore, management, like any other art, requires a 'knowledge-based practice' and 'its application'. It is doing things in a situation in the light of realities. Therefore, Follet defined it as *doing things in the light of realities of a situation* and thus managing is an art similar to other practices like music, medicine, engineering etc.

Terry pointed out that, *the art of managing advances with practice, through feeling, imagining, describing, thinking and creating*. This holds true for recognised arts like poetry, painting and interior decoration. For example, interior decoration is a creative art of adjusting space and equipment with two aims, i.e. order and beauty, so that it is performed to suit the cultural and social needs of the dwellers. Similarly, management is an art for achieving the desired goal adjusting and allocating the use of the available resources.

Another person to relate management with art is Luther Gullick, who defined management as a '*field of study*' that seeks to systematic understanding of why and how men work together to accomplish their objectives. He thereby stressed the point that management has to be based on knowledge, understanding and then following it in a systematic way while working together. According to him, management meets the requirements for a field because it has been studied for sometime and has been organised into a series of theories. He holds the view that the field of management would truly become a science, when theory would be able to guide managers by telling them what to do in a particular situation, to enable them to predict the consequences of their actions.

Henry M. Boettinger is also of the opinion that *management is an art*. Like any other literary or a fine art like poetry or painting, management requires the same three components - the artist's vision, knowledge of crafts and successful communication. In all these respects, management is an art. And therefore, just as artistic skill can be developed through training and practice, managerial skills (i.e. art) can be developed in the similar way. For example, in the home, a homemaker can develop the skill of planning and purchasing of equipment and furnishings for the home through learning and practice. Art is thus, the most creative of all human pursuits and so is management.

This is again emphasized by Koontz O' Donell and Wielrich who have pointed out that, *when the importance of effective and efficient group cooperation in any society is appreciated, it is not difficult to argue that managing is the most important of all arts.*

Although, it is generally argued by a number of persons, that management is an art, it is universally not agreed upon. Managing as a practice is an art, the organised knowledge underlying the practice may be referred to as a science, as the most productive art is always based on the understanding of the 'science' underlying it. In this context, science and art are not mentally exclusive but are complementary.

In the case of a practical science like medicine, measurement of normal and abnormal health shows a state of well-being in a state of comparative scale. Similarly management is measured or judged as 'good' or 'poor' on a scale or degree of competence. Analysis of business failure has shown that a very high percentage of them have been due to unqualified, incompetent or inexperienced management. When these enterprises are managed well, they will be able to succeed. This can also be held true in the homes, where many goals are left unfulfilled if not managed properly. However, if an efficient homemaker is able to manage her home well without undue hardships, it may result in complete satisfaction to all the members. Thus, a well-managed house will find the family members happy and satisfied as it is likely that their goals are achieved.

The study and analysis of management have lagged behind other sciences until recent years. The essential feature of science is that, the knowledge has been discovered and systematized through scientific methods. Science is a systematized body of knowledge pertaining to an area of study and contains some general truths explaining past events or phenomena. In management, science provides a body of principles or laws for guidance in the solution of specific management problems and in objective evaluation of results. The analysis of basic management functions has led to the development of certain principles, which can be applied as general guides for solving concrete problems in the future. For example, **budgeting** is accepted as the first principle of planning financial management of all organisations, be it a highly organised and formal sector like a corporate business or a less organised informal organisation like a home.

On the other hand, the art of management deals with the application of skill and effort for producing desirable results or situations in specific cases. To be effective, the art of management must be grounded in the knowledge of principles. Fundamental principles always react upon the art and shape the way of doing specific things. That is, the science and art are interrelated and complementary. With every increase in the knowledge of science, the art is bound to improve. "Management science and management art are thus interwoven and overlapping in nature."

The figure 1.3 sums up the characteristics of management as science and as art. Therefore, we can conclude that even though some aspects of management have become scientific, still a major part of management remains as an art.

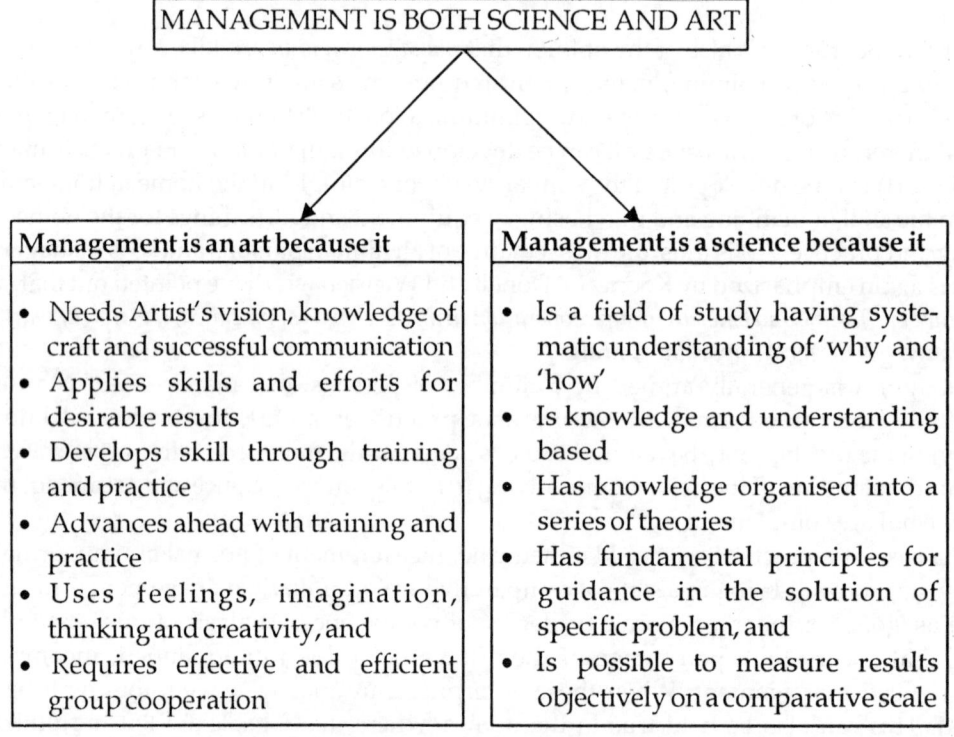

Fig. 1.3: Management as Science and Art

MANAGEMENT AS A PROFESSION

An analysis on the nature of management shows that it is 'partially art' and 'partially science'. As the study of management has advanced further and because of its inevitable role in all organisational activities, another question arises now, "is management a profession?"

Edgar H. Schien[6] has compared key qualities of professionals with those of managers. The three characteristics as pointed out by him need a mention here.

---

[6]Schien, E. H. (1986). *Organisational Culture and Leadership.* London; Jossey Bass Publishers.

First, professionals base their decisions on **general principles**. There are certain reliable management principles, which are shown by the very existence of management courses and training programmes. Although, the principles held in common by most managers and management theorists do not apply perfectly in all situations, particular guidelines have high reliability.

Second, professionals achieve status through **performance** and not through favouritism, or any other factors irrelevant to work at hand. However, it happens at times, though unfortunately that some people achieve managerial positions through their relationships or because of their economic power. Besides, there exists no body who can help in setting standards of judgement in the field. Because of the complexity of factors that enter into a managers' role, it is more difficult to judge managers, than the people engaged in professions like medicines or law, as surgeons or lawyers, respectively.

Thirdly, Schien points out that professions must be governed by a 'strict code of ethics' that protects their clients according to the expert knowledge in specific areas. Their clients are dependent upon them and as a result, are in a vulnerable position. But, unfortunately, management code of ethics is yet to be developed. Schien concluded by saying that by some criteria, management is indeed a profession, but by other criteria it is not. Today management seems to be emerging as a new field of professionalism in all kinds of organisations. Current social pressures seem to be bringing about a heightened awareness of ethical standards. The growth of formal management teachings and skills, through reputed IIMs (Indian Institute of Management) and universities, addressing the ethical issues, are the hallmark of profession.

Another characteristic of professionalism as suggested by Borae O Sale Birg is 'dedication and commitment'. In the kind of true professionalism, it is necessary to combine life and work through personal dedication and commitment. According to this criterion, managers can also be counted as professionals in the best sense of the word.

Thus it is clear from the above discussion that management, besides being a science and art, is also a profession as shown in Fig. 1.4. Though men primarily dominate this profession, women have also marked their entry and have proved to be themselves to be as good as men. Let us now look at the present scenario and role of women as managers in it.

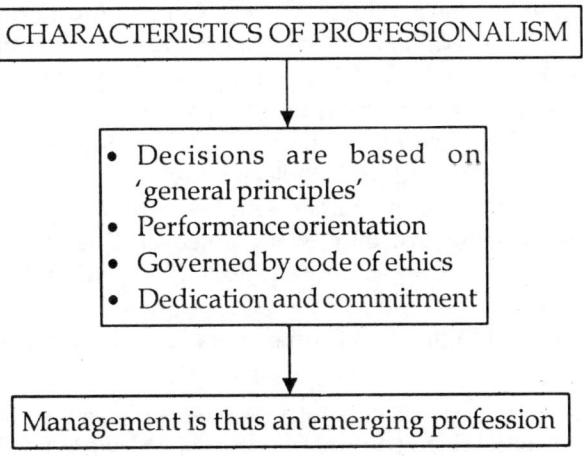

Fig. 1.4: Management as a Profession

WOMEN IN MANAGEMENT

The science and art of management is being extensively used by women in every field of life, including that of a home. Traditionally the managerial skills of women were restricted to home and its activities. Though it is treated as an informal sector, managing the home had always been a big challenge for every woman, as she had to toil the whole day to be successful in her role. The present day home based managerial activities pose bigger challenges, as her role is now extended beyond the four walls of the home. However, management of home for centuries have given women managerial skills, which enabled them to become successful managers of bigger and more formal organisations as well, although, home management is still their prime responsibility. Thus, women of today have dual responsibility of management at home as well as in the organisation in which they work.

The advent of the industrial revolution and consequent technological developments coupled with advanced education and changing socio- economic needs brought about enormous changes in women's lives and their roles. While job opportunities in both public and private sectors have increased, more and more women are using their managerial skills at work, which were earlier confined to the management of their homes alone.

While careers of women have already advanced greatly in western countries, in India, the scope of women's career has expanded only in the last forty years or so. The Indian Census (1971, 1981, and 1991) reports[7] that overall Work Participation Rate (WPR) for women has increased from 14.22% in 1971, 19.67% in 1981 to 22.27% in 1991. The employment and unemployment in India 1999-2000 report[8] shows that it has increased to 25.9% by the year 2000. Women's employment[9] in the organized sector has increased significantly from 1.9 million in 1971 to 12.6 million in 2001. Of these 2.6 million women (62%) were employed in public sector and 1.8 million (38%) in the private sector. The most spectacular increase has been in the number of Indian women working in secretarial, academic, managerial and administrative capacities. According to 1991 census report, the professions, which have gained most popularity among women, include areas related to administration and management. This trend was also confirmed in census 2001 report.

Business Today[10], identified the 25 most powerful women in Indian business. Some of the Indian women who have secured a firm place for themselves in the managerial world today are Vineeta Rai, country's first woman Revenue Secretary, Kishori J. Udeshi, Deputy Governor Reserve Bank of India, Shikha Sharma, Lalita Gupte, Kalpana Morparia, Renuka Ramnath and Chanda Kochhar of ICICI Bank, Naina Lal Kidwai of HSBC securities and capital market. Some others in this list are Rashmi Paliwal, a Delhi based exporter, Sarparveen Kaur, a Jalandhar based honey producer, Sunita Pitamber of Bombay, Sunita Narain director of Centre for science and environment, Amrita Patel chairperson of National Dairy Development Board, Kiran Mazumdar Shaw, chairperson of Biocon India, country's Biotech Lady.

Similar trends showing increase in women employment and participation in managerial activities are reported worldwide. Powell (1988) pointed out that in United States of America (USA), 38% of women were working as nonfarm executives, administrators and managers in 1987, as compared to only 4% in 1900 and 16% in 1970. She quoted the survey findings of American Management Association (AMA) that female managers are more committed to their career than

---

[7]Govt. of India. (1991). Census of India: Population Totals: Workers and their Distribution. New Delhi.
[8]Employment and Unemployment in India (1999-2000). NSSO Report No. 548. (55/10/2).
[9]Census of India 2001: Population Profiles, Office of Registrar General, New Delhi, India.
[10]The 25 Most Powerful Women in Indian Business (2000, Nov. 23). *Business Today*.

male managers of equivalent age, education, salary and managerial level. To hold managerial positions, women have greater barriers to overcome than men, both in their internal predisposition and in others' expectations from them. A female manager with a position and background equivalent to a male manager needs to have a greater commitment to her career to overcome sex difference barriers. The research done to find out the differences in male and female managers concludes that they differ in some ways but for the most part, they do not. Sex differences have generally not been found significant in global measures of management behaviour. The differences in their responses, in fact, favour women managers as they are more committed, achievement oriented and perfectionists. The myth that men make better managers is simply not true in the present context. Research studies have thus proved that women make as good managers as men, or perhaps, even better.

Today women also have entered into this profession of management while continuing to manage their home. They are as successful managers of business organisations as they are home managers. With the advancements in the concept of management it must be remembered that amongst the important things concerning management, it is a means of adjusting to change in an orderly manner. These changes occur both within the family as well as in the external environment.

ETHICS IN MANAGEMENT OF FAMILY LIVING

Ethic is a science of morals. It is that branch of philosophy, which is concerned with human character or conduct. Ethics are also professional standards of conduct in accordance with the approved moral behaviour. Therefore, they are the guiding principles, which help an individual in selecting the moral behaviour. The concept of ethics in management is emerging as an important concept, because of its relevance today. In recent years, management is accepted and recognised as a profession. Thus, the need of a code of conduct is also felt more than ever before. In this chapter we are trying to address some of the important ethical issues in management with special reference to family living.

- **Unbiased Decision:** It is important for a manager to take decisions based on needs and situation. The need for unbiased decisions is the most important ethical issue in management. More often than not, decisions are based on favouritism, and individual whims and preferences. It is important for a manager to avoid biased decisions and to take democratic decisions to improve performance, efficiency and efficacy. These qualities are more important to an organisation than individuals who are not likely to follow these principles. An unbiased and democratic manager can earn both respect and trust from others without sacrificing the results or goals. She will prove herself to be "right" in the long run.

- **No Gender Bias:** A differentiation among individuals on the basis of gender is the most unethical practice. Boys and girls should be treated similarly. Giving more preference to either is not acceptable. A manager should provide equal opportunities, resources, and motivation to all the members of the family irrespective of their gender. In India, gender bias in families is more common and prevalent. This needs to be addressed as an important ethical issue. Negligence of a girl child or a woman in a family give rises to unethical behaviour like subservient status, violence towards them etc. It has also led to unethical social practices like dowry harassment and deaths, female foeticide and female infanticide. Thus, it is for a homemaker to ensure that all males and females are treated equally in a family.

- **Respect for Tradition and Culture:** We have already discussed that management has a culture bias even though it is universal. Thus, every family follows its own customs and traditions, which also form their personal identity. However, it is very important that one family should respect other's customs and traditions as well. Respect for your own family's traditional cultural practices is as important as respecting cultural differences among families. Thus, a manager should follow their traditions and customs with pride and simultaneously pass them to the younger members of the family, because they give directions to life.

- **Ethical Attitude:** Ethical attitude includes those simple day-to-day behaviour, which makes family and social living a pleasure. Waiting for your turn and following rules can make community living problem-free. More often we see that traffic jams are caused because of someone's impatience to follow traffic rules or their inability to wait for their turn. Similarly if we follow simple rule which are made to make life easy rather than evade them, living together could become problem-free. For this, one must also respect others' needs. As family is a child's first school, these attitudes are acquired and formed at home. Ethical attitudes can be developed by simply abiding laws and meeting legal requisites so that children can observe them and internalise these behaviours as their positive attitudes.

- **Community and Social Living:** Man is a social animal. He lives in a society which has people with diverse view and practices. Harmony in community and social living can be achieved only by ethical behaviour. But we have to first recognise the need and importance of community and social living. We are dependent on society and our community for our growth and development as well. We are an integral part of society. Thus, it is important to follow ethical behaviour for happy living. A home-manager's (the manager with her team of her family members) contributions can ensure social and community well being and harmony. This is specially true in the days of social unrest.

- **Social Responsibility:** As a part of society we have few social responsibilities. All of us use community facilities like roads, commercial areas, parks etc. Therefore, it is our responsibility to raise voice against misuse and encroachment of public land. Public land is often encroached for making houses, growing gardens in front of houses or for raising cattle or domestic animals or for parking vehicles. Also, sometimes thrash is thrown on public roads. By rectifying these behaviour or simply creating awareness and a sense of social responsibility among the family members and others, one can influence and create a sense of social responsibility among them.

A home-manager can help every member of her family in leading an ethical and happy life. Thus, these issues are very important in realising not only the goals for family but also for the harmonious social and community living, to make the world a place for worthy happy living.

2

Managers and Management <<<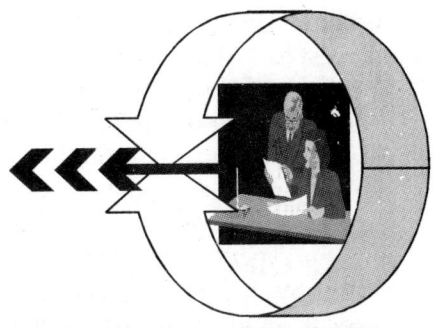

THE role and importance of management in today's life is widely accepted now. However, the success of the process of management depends entirely on its manager. Manager is the person who is responsible for carrying out this process. He/she is the one who manages the organisation to achieve its goals by using the knowledge and the principles of management. Thus, the success of the management and the achievement of goals depend upon the managers. Similarly, the complete satisfaction of family members depends on the homemaker, who ensures the achievement of all family goals. Let us now look at the roles, responsibilities and characteristics, qualities, and types and kinds of managers in an organisation, as managers are important in an organisation and they can make it a successful one.

MANAGERS AND ORGANISATIONS

How successful an organisation is in achieving its objectives and in meeting its needs depend upon how well the organisation's managers do their job. If the managers do not do their jobs well, the organisation will fail to achieve its goals. Just as managers function within the organisation, organisations function within the larger society. The performance of its organisation as a group is a key factor in the performance of a society or a nation, which again will be reflected by the performance of the organisation. In the context of a family, a homemaker while managing the house, regulates every members' performance, thereby regulating the performance of the entire family as an organised unit.

Peter Drucker[1], one of the most respected writers on management, argued that a manager's performance could be measured in terms of two concepts - efficiency and effectiveness. As he puts it, efficiency means 'doing things right', and effectiveness means 'doing the right things'. As he points out, one of the most important qualities of a successful manager is his/her efficiency, i.e. the ability to get things done correctly is an 'input-output' concept. An efficient manager is one who

---

[1]Drucker, P. (1977). *People and Performance*. New Delhi; Allied Publishers Pvt. Ltd.

achieves output or result that measures upto the inputs such as labour (human resource), material and time which are utilized to achieve them. Managers who are able to minimize the utilization and cost of the resources to attain their goals, are considered efficient.

But it is not enough to measure the efficiency of a manager's use of resources only in terms of keeping their costs low. As people are the major resources for most of the organisations, efficiency might also mean developing or upgrading the skills of the people who work in the organisation.

Homemaker acts a manager in the family. Her performance affects the family similar to that of a manager in any other organisation. Thus, an efficient homemaker should attain family goals, with the expenditure of least amount of family income, family assets, time and energy of self and that of the family members. While doing so, she should also ensure personal development, happiness and complete satisfaction to all her family members.

The next important quality of the managers is being effective. Effectiveness, here refers to the ability to choose appropriate objectives. An effective manager is one who selects the right thing to get the work done completely. For example, a manager in the home who selects an inappropriate objective, such as, extra clothing for children when the demand for extra expenditure in education is required, is not an effective manager. Such a manager would be ineffective even if she chooses the best quality and well fitting dresses which are the indicators of her efficiency in selecting these. No amount of efficiency can compensate for the lack of effectiveness. For an efficient manager, it is also necessary to be effective.

A manager's responsibilities thus require a performance that is both efficient and effective. Effectiveness is the key to the success of an organisation and the managers need to make the most of the opportunities in dealing with the organisational needs. In this way, the responsibility of the manager is more important for the organisational success. Thus, a homemaker is a key resource person, whose successful performance determines the extent of satisfaction and happiness for her family.

ROLES OF A MANAGER

Henry Mintzberg points out that there is a considerable similarity in the behaviour of managers at all levels. All managers, he argued, have formal authority over their own organisational units, and they desire status from that authority. This status causes all managers to be involved in interpersonal relations with subordinates, peers and superiors, who in turn provide managers with the information they need, to make the decisions. These different aspects of a manager's job cause managers at all levels to be involved in a series of interpersonal, informational and decisional roles, which Mintzberg termed as 'organised set of behaviour'. The following discussion is focussed on these roles.

Interpersonal Roles

Three interpersonal roles help the manager to keep the organisation running smoothly. Although, the duties associated with these roles are often routine, the manager cannot ignore them, as they are to be performed without fail.

The first interpersonal role is that of **figure head**. As the head of a unit, the manager sometimes acts as a figure head by performing certain ceremonial duties — greeting visitors, attending subordinate's or relative's wedding, etc. Second the manager must play the interpersonal role of **liaison** by dealing with people other than subordinate/staff or superiors/family relatives, such as peers within the organisation and suppliers or clients outside of it. Finally, the manager adopts the **leadership** role—hiring and training employees/domestic help, and motivating and encouraging family members.

Informational Role

There are three informational roles in which managers gather and disseminate information. The first is the **monitor role**. As monitor, a manager constantly looks for information that can be used for advantage. Subordinates are questioned and their answers are gathered as information. Unsolicited information is also collected usually through the manager's system of personal contacts. The second, is the **disseminator** role, and the manager distributes to the subordinates the important information that would otherwise be inaccessible to them. Finally, as a **spokesman,** the manager transmits some of the information outside the unit or even outside the organisation. Keeping the superiors in the organisation satisfied by keeping them well informed, is also an important role of the managers.

Decisional Roles

There are four decisional roles the manger adopts in an organisation.

In the role of an **entrepreneur,** the manager tries and decides to improve the unit. As an entrepreneur, the manager initiates all activities voluntarily.

In the role of a **disturbance handler,** the manager responds to the situations that are beyond his/her control, by taking certain decisions.

As a **resource allocator,** the manager is responsible for deciding how and to whom the resources of the organisation and the manager's own time will be allocated. Also the manager screens all the important decisions made by others in the unit before they are put into effect.

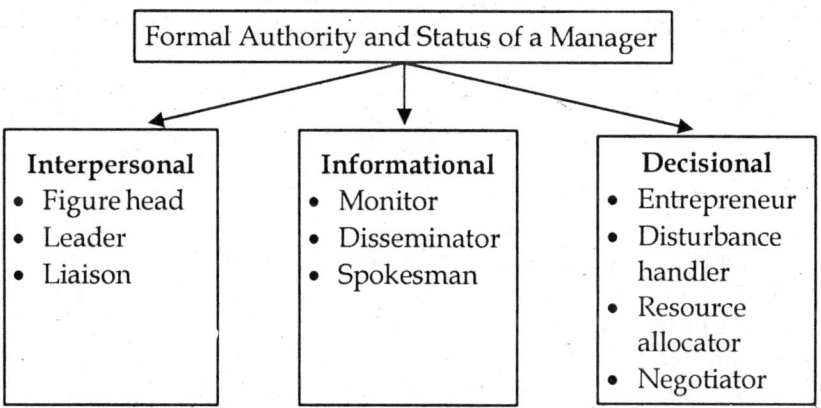

Fig. 2.1: Expected Roles of a Manager

The fourth and the last decisional role is that of a **negotiator.** In a formal organisation, managers spend a great deal of their time as negotiators, because they only have the information and authority that negotiators require, such as a company president who works out a deal with a consulting firm or a production head who draws up contract with a supplier and so on. However, at home, homemaker may also need to negotiate while making purchases for the home specially in the purchase of major household equipments. The diagrammatic sketch (Fig. 2.1) very clearly indicates all these roles of a manager.

ROLES AND RESPONSIBILITIES OF A HOME MAKER

As pointed out earlier, management is required not only in business but also in various home activities. Thus, it is important to study the role of a homemaker, as she is the one who acts as a manager in the home. To be an effective manager, the homemaker must possess special qualities for being effective and efficient. This is specially important because she wants to run the home smoothly. An efficient homemaker will have the ability to get things done correctly. She need to know the optimum utilisation of resources both human as well as non-human ones. Following discussion focuses on the special roles of the homemaker as interpersonal, informational and decisional.

Interpersonal Roles of a Home Maker

The various interpersonal roles of a homemaker as the manager in the home include the following:

- Guiding the family towards development of a sound philosophy of life.
- Channelising personal relationships which are wholesome and satisfying.
- Taking care of children and their needs such as feeding, bathing, cleaning, clothing etc.
- Inculcate ethical habits and qualities among the family members such as tolerance, patience, cooperation, harmony, love and sympathy, and develop satisfying relationship within and outside the family.
- Appreciation of differences as means of enriching life, and an allround personality development.
- Planning for participation of family members in community activities as responsible citizens.
- Acceptance of mutual responsibility for family and community.

Informational Roles of Home Manager

The informational roles of the manager in the home are

- Developing a sound philosophy with sustaining values.
- Planning for and providing nutritious food and suitable housing and clothing for the family.
- Planning for and helping to maintain health of the members of the family by keeping the knowledge of the human body, mental health, family planning, formation of good habits and care of the sick and the old people and of first aid which can be utilised during emergencies.
- Planning for and guiding the educational and social development of the members of the family.
- Making the family members aware of the various 'public facilities' or community resources such as the water supply, electricity, cooking gas, telephones, hospitals, transport, public parks, library etc., and their judicious use.
- Giving knowledge about the various local services provided to them in case of emergencies, such as fire, theft, accident, death etc.
- Disseminating the information about the various social organisations like Resident Welfare Association (RWA), Mahila Mandals, Mahila Samitis, Yuvak Kendras, Panchyats and community centers.

DECISIONAL ROLES

The manager in the home is expected to perform the following decisional roles also:

- Setting family goals.
- Planning the family members' time and energy resources in such a ways that the work is done and their demands are satisfied.
- Planning and guiding the use of family income, bringing a balance between income and expenditure, supplementing family income, besides saving money for the future etc.
- Planning for and providing for the family, which will meet the day-to-day needs of its members.
- Planning for purchasing equipment and furnishings for the home.
- Being aware of her rights and responsibilities as a consumer. This shall enable her to use her knowledge to choose the best of her ability to select the right quality product in the right quantity at the right time for the right price.

The managerial responsibilities as mentioned so far are all interrelated. For example, family finances and their management touch all phases of life as they influence the desires, decisions and choices of all the family members. To a great extent, they control what the family members can have and do throughout their life. Also planning for clothing the family is closely related to family finances. Similarly, management of time and work is also closely related to other managerial responsibilities. To be able to accomplish the daily household work without undue strain and tension, the housewife should think of each managerial responsibility and task to relate to it in terms of time and effort required for their accomplishment. However, a manager can successfully fulfill these roles and responsibilities only if he/she possess certain qualities and characteristics.

QUALITIES OF A HOMEMAKER

These are those inherent abilities, which enable a manager in discharging his duties as expected of them. It is these qualities which underline the statement that 'managers are born'. However, management is both **caught** and **taught**. However, for mastering the science and art of management, the following qualities are essential.

- **Intelligence:** An intelligent manager is keen on observing, understanding, thinking and remembering. She should develop all ways of gaining knowledge, ability to solve problems and achieve goals. Intelligence also includes mental alertness and ability to grasp knowledge and situations. She should be able to see the situation as a whole and analyse the situation through the relationship between the old and the new.

- **Enthusiasm:** This is a matter of temperament. It is an ability to motivate others to become interested in the activity. The environment created should be contagious and make everyone to take interest in performing and completing the work. An enthusiastic manager can create a feeling of happiness by working hard to achieve her targets. Often the enthusiasm of a manager is sufficient enough to motivate and lead her team or family members towards their goals.

- **Imagination:** This is the ability to rearrange facts and ideas into new patterns of work. This aids in making fresh plans and come out successfully in meeting problem situations. This also helps in bring out novel plans, add variety and takes out monotony in work situations. It can also be called creativity. Just as creativity and imagination are important ingredients to all forms of art, so are they for management.

- **Perseverance:** A good manager should have the ability to combine courage and patience to face facts. She has to believe in the inherent values of the task and feel strong about the outcome of the efforts. Only a confident manager can make her team or family members believe in her as well as in the tasks they are doing. Faith in work and abilities will give the manager the courage to take necessary risks and try to achieve solutions until the problems are solved and goals are achieved.

- **Judgement:** A sense of judgement enables the manager to be fair in weighing the various facts in a situation and to see the problem in totality. She should be able to weigh critically and make decisions while choosing the best alternative. Decision making, being the crux of management, stresses upon this quality of a manager. As the level of manager goes higher in the hierarchy, she has to take more and more decisions, which are more critical as well. Simply stated, a good judge is a better manager.

- **Adaptability:** Flexibility and adaptability are synonymous terms. The human environment and the work demands are not static and require change in the approaches. All managers have to adapt themselves and their approaches as and when required. Also it is not possible for a manager to predict the future with absolute certainty as many things in the environment are beyond her control. Plans need to be adapted according to the situation at the time of its execution. Therefore a manager should also keep derivative plans or alternative course of action in-built in the master plan well in advance. An intelligent manager will be able to adapt quickly in changing situations and changing demands.

- **Communication:** Ability to communicate one's views and ideas is an important quality of a manager. A manger should be able to convey clearly and have a meaningful dialogue with the colleagues or family members. This not only involves sharing of knowledge, feelings, desires and experiences, but also taking time to share thoughts with one another. An effective communication is the key to managers' success. With good communication skills, they can build visions, create dreams and clarify expectations regarding future and divert efforts to achieve them.

- **Understanding Human Nature for Team Building and Leadership:** The managers should be sympathetic and understanding since these are important in developing healthy human relationships and for reducing friction while dealing with others. If management is a process of getting things done by the others, a manager has to be a psychologist, a sociologist, a philosopher and a mentor, all at the same point. The manager has to understand the values, attitudes, behaviour and capabilities of the team or family members so that she is able to delegate the right responsibility to the right person and lead and motivate them towards the achievement of goals. It also helps a manager in selecting, building and leading teams and enhancing their skills for greater achievement of goals.

- **Self-Confidence and Self-Esteem:** They are important qualities of a manager which helps in decision-making and leading the whole family towards their goals. A confident person with the faith in herself and her team can easily lead the family towards success.

- **Synergy:** A manager can use synergy by combining and coordinating the individual capacities to achieve higher total output. When human efforts are combined, the total efforts are much more than the addition of their individual efforts. This is synergy. This quality helps a manager in leading the team effectively and efficiently.

- **Self-Management:** It involves the management of one's own feelings. Home is a highly personalised enterprise with intimacy and affection. It requires personal adjustments to suit all its members and develop a relationship to minimise conflicts. A manager has to set example by adjusting to people and situations and reorganise herself as the situation demands.

CHARACTERISTICS OF A HOMEMAKER

The eminent management expert, Fayol lists the following characteristics of a manager. These characteristics help her in doing her job effectively. These characteristics are more apparent and help a person to easily recognise and distinguish herself as a good manager. The criteria also act as guidelines for selecting people as managers in an organisation, though, in a home situation, a homemaker has to make efforts to develop these characteristics. Thus, they form important guidelines for efficient and effective managers.

- **Physical Health:** A manager has to a have a good physical health and vigour to perform the duties properly. In the absence of a good physical health, the manager is not in a position to carry out the work successfully. A healthy mind lives in a healthy body. Therefore, good physical health is essential for developing good mental health as well.

- **Mental Health:** Similar to good physical health, it is important to have good mental health too. It provides the manager the ability to learn, understand and judge in logical and unbiased manner. A good mental health also aids in adopting oneself in changing situations.

- **Moral Responsibility:** This includes the characteristics such as being firm, willing, loyal, dignified and factual. The other moral values like teamwork, sincerity, honesty and justice are of equal importance as they help managers in securing the faith of their boss, subordinates or family members.

- **Educational Qualification:** A manager needs to be knowledgeable and informative. She should generally be aware of all that is going on in the workplace. Education helps and trains her in performing her roles as monitor and disseminator of useful information and as spokesman for her organisation. An educated person can also allocate resources better and can be a better negotiator than a one with lesser education.

- **Technical Knowledge:** A good technical knowledge is an important aspect of management. A manager needs to know all about the functional and operational areas of her organisation. It is essential for a manager to be equipped with the latest technical knowledge as it gives her an extra edge in her field. Also a manager with thorough technical knowledge can monitor and supervise better. She can also work out newer and better technical solution to problems by developing new alternatives. As management has become a specialised field, managers of each field must have knowledge of her field i.e. a finance manager must know about all technical aspect of finances and similarly, a sales and advertising manager should have knowledge of media, communication skills and a home manager should have full knowledge of managing family finances, child care, meal management, clothing etc.

- **Work Experience:** Experience is richer than qualifications. While an educational qualification provides a theoretical knowledge, experience provides her the practical solutions, thus enabling her to face and tackle challenges and come out of it successfully. Experience to some extent can make up for the lack of education. It acts as self-educator.

Thus it can be seen that a manager need to be a master as well as a jack of all trades. The success of a manager depends entirely upon how well and quickly she develops these qualities. These qualities and characteristics are of great help to a manager in performing her function. Now let us have a look at various functions of a manager in any organisation may it be business or home.

FUNCTIONS OF A MANAGER

In any organisations there are some functions, which every manager has to perform. These functions are integral part of any management process. If a person wishes to apply the process of management to any activity, he or she will have to perform these functions.

- **Decision Making:** It is an integral part of any management process. A manager has to take series of decisions from the very beginning till the end of the process or till the time the goals are achieved. Decision-making is also called as the crux of management. Each role or every other function involves decision-making. As a manager gains higher position in the organisation, the other roles and functions are replaced by the decision-making function.

- **Planning:** Planning is a process of pre-determining the course of action. It involves premising the future and selecting the best course of action in advance. Every manager has to decide well in advance what is to be done, how and by whom. This, is done by setting goals and directing efforts and allocating resources for the achievement of pre-determined goals. Planning is also said to bridge the gap between, where we are and where we want to be and this is a major function of a manager of every activity.

- **Organising:** This function of a manager involves classifying the activities listed in the plan and delegating them to the people who would best be able to do them. The manager has to identify the appropriate human resources and allocate tasks to them. This function implements the planning that had been done earlier. Success of any plan depends upon how well it is organised and executed. In this function a manager distributes not only responsibilities but also suitable authority as well to the subordinate or family member to carry out the tasks in the best possible manner. In a way. the manager sets up a team where each one has a specific work and a specific role to play. This team organisation ensures easy, smooth and continuos flow of information and work together towards the desired goal.

- **Staffing:** This function of a manager involves human resource management. In a formal organisation, requitement and training are the biggest human resource challenge faced by a manager. Selecting the right person for a job and training the existing staff for their personal and professional growth are a must for the organisational development. This is again a major function of a manager, who creates an organisational structure in the organisation. Organisational structure in turn defines the level of each person, their roles and responsibilities, authority relations, information flow and remuneration packages for every one in the organisation. At home, a homemaker though do not select a family member, but perhaps does select a domestic help. She also trains family members to carry various household tasks and ensures their both personal and professional growth.

- **Delegation and Guidance:** In any organisation a manager has to delegate responsibility for doing various tasks and necessary authority to enable them to successfully execute these assigned duties. Delegation requires an in-depth understanding of capabilities, strengths and shortcomings of a person, chosen for a task. After selecting a person for carrying out a particular task, a manager will have to guide or direct that person to complete that task. This includes giving clear instructions and stating as clearly as possible, the expected outcome. While guiding a person to do a task, the focus is not only on the necessary work outcome but also on the personal development of the person carrying out that task.

- **Leading:** In this function, a manager has to gain co-operation of the subordinate and the family members and motivate them towards the goals. To lead, a manager must first understand the attitudes, philosophy, values and behaviour of each member of the organisation or a family. Then she will influence them to shape a desirable social system within the organisation or the family. The manager has to do this by guiding them and motivating them to improve their organisational effectiveness. The manager is also responsible for creating favourable environment for performance. While doing so she will establish relationship between self and subordinate and family members. An effective and efficient leader will be a successful manager as well.

- **Controlling:** Controlling is correcting deviations on time. This function of a manager involves identification of problems and discrepancies from the plan of action and taking measures on time to correct them. It involves measuring performance against pre-set goals. An effective manager builds control standards while planning and clearly states them to the subordinate or family members. This also acts as guidelines for subordinate or family members and acts as an appraisal system for the manager. The control function is the most effective way of cutting costs and losses incurred by errors. Thus every manager uses certain control devices like budgeting, inspections, record keeping and maintaining attendance etc., while performing the control function. Control is essential since the activities are pre-planned, and the activities are to be performed according to the plan.

Though all managers perform these functions, they can be differentiated by the way they perform these functions. Let us now have a look at the various types of managers found in the organisations.

TYPES OF MANAGERS

It is the responsibility of a manger to lead the group in a right direction for the achievement of the pre-set goals of the organisation. Thus people and work are the most important ingredients, which signifies the styles adopted by the manager to execute the functions. On the basis of boss-subordinate/homemaker-family member relationship, and the concern for people and work, managers can be classified into various categories. The four basic types of decision-making behaviour are identified among managers, which clearly state the manager-subordinate or homemaker-family member relationships. Thus managers can adopt four styles which will determine their relationship with others in the organisation.

➤ **Autocratic ("tells") Style:** In this style a manager usually makes the decisions promptly and communicates them to the subordinate or family members. The manager expects them to carry out the decisions loyally and without raising difficulties. Therefore, this style emphasis upon 'telling' others what is to be done. The information flow is downwards and the communication is carried out from top to bottom alone. The manager adopting this type of style is very confident about the decisions and does not care how much others think or believe in the issue.

➤ **Paternalistic/Persuasive ("sells") Style:** In this style also, a manager makes decisions promptly but before going ahead, tries to explain them fully to the subordinate family members. The manager gives them the reasons for decisions and answers whatever questions they might have. Even though the manager takes the decisions, she tries to convince others by answering their queries and explaining the reasons to them. Thus the emphasis is on selling the decisions to others. The information flow in this case is also downwards. However, there is some communication that takes place between the manager and the subordinate or family members. This style shows that a manager, though is very confident about the decision, would also like to gain the confidence of the subordinate or family members.

➤ **Consultative ("consults") Style:** In this style, a manager usually consults with the subordinate/family members before reaching decisions. Later she announces the decisions and expects everyone to work loyally to implement it, whether or not it is in accordance with the given advice. The emphasis in this style is upon two way communication and information flow as manager consciously seeks the information and opinion from the subordinate or family members. However, she keeps the final authority of decision making with herself. This style can be interpreted as interdependence between the superior and the subordinate or homemaker and family members. In a nutshell, manager believes in making decisions in consultation with the others. It also happens in cases where she is not very sure or confident about the task to be decided.

➤ **Democratic/Participative (Majority Vote, "Consensus") Style:** In this style, a manager usually calls a meeting of the subordinate or family members when there is an important decision to be taken. She then puts the problem before the group and invites discussion. In the end she accepts the majority viewpoint as the decision. The information flow is upwards and the power to decide rests with the majority. In this style, subordinate or family members tend to polarize into dependence and counter dependence. Some times wrong decisions can also be taken simply because more number of people believe in it. The qualities of a manager and her competence do not come in light in this style. Managers usually adopt this style when they either do not know much about the issue or when the issue is very sensitive and managers do not want to take the accountability for the decision. They rather reduce the risk by working with popular opinion. In this style, subordinate and family members are usually very happy, as they feel they have taken a decision themselves.

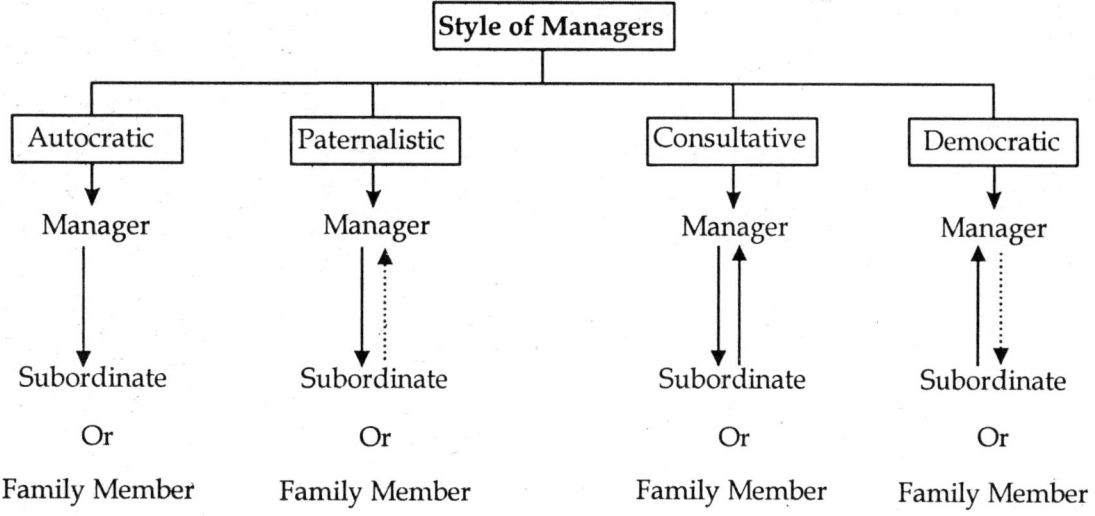

Fig. 2.3: Different Styles of Managers

The figure 2.3 indicates the styles adopted by managers. Thus, we can conclude that decision making styles influence manager-subordinate or homemaker-family member relationships. A manager may not follow any one of the styles alone all the times. She uses a combination of these styles. The situation, environment, importance and sensitivity of matter, technical knowledge and experience of a manager about the issue, collectively govern the decision making style to be adopted. Similarly, managers can also be classified on the basis of their concern for work or people. On such basis, there are two types of managers. They are:

- People centered, and
- Task centered

People Centered: They are also called humanitarian managers. In an organisation, such a manager deeply cares about her people. The decisions are governed by the concern for the well being of the others. The manager gives little importance to work. The concern is more about the welfare of employees or family members than the accomplishment of the tasks. The manager gives complete attention to the needs of the staff and family members. The subordinate or the family members are usually very happy but it is likely not much work is done or the work is done with little efficiency.

Task Centered: This manager has a task focus, and believes in efficiency at all costs. Performance is important to her and treats others according to their efficiency and performance. The manager is more concerned about the work than the people. Also, she gives little thought to the needs of the employees and family members. The focus is on the task to be accomplished. The subordinates and family members are usually strained, stressed and not very happy. However, best performance is ensured at least costs. This type of manager is usually not very popular among subordinate or family members but gets the best results for the organisation.

A manager cannot be completely people centered or task centered. However, she is a combination of both but can take a tilt towards either side. Thus every manager continuously tries to strike a balance between people-focus and task -focus.

An efficient homemaker must possess special qualities for being effective and efficient. She should know how to have optimum utilisation of the available family resources both human as well as non-human. As a manager, she has to perform many functions in the family like decision-making, planning, organising, staffing, leading and controlling. While performing these functions, she may adopt certain styles. A manager can be autocratic, persuasive, consultative or democratic, and can also be people-centered or task-centered. Usually a manager uses a combination of these styles depending upon the situation and the environment.

3

Management—Principles and Process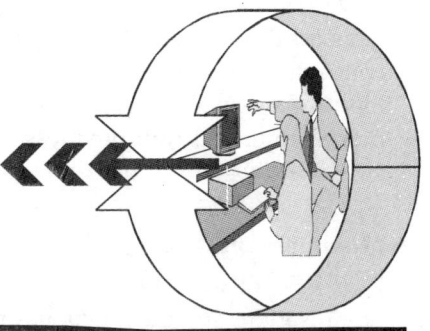

FOR effective management, the manager has to play certain roles. As a homemaker and the manager of the home, she has some responsibilities and has to play some important functions, for which she needs to possess some qualities. However, without understanding the management principles and processes, the management becomes incomplete. Majority of references to management is from corporate or private sector industry, probably because the greatest noise and initiative related to the development of management skills for achieving greater success and profit initially came from such sources. However, management principles and process are used in running every organisational system — whether a family, sports club, business, school, college or the entire nation itself. Thus, this knowledge of management relates to the decision centers of all organisational systems. This chapter mainly centres around the application of the knowledge of management principles and process given by eminent thinkers to the management. Let us first see how management emerged as a separate discipline and its growth in general.

HISTORY AND GROWTH OF MANAGEMENT

Though management emerged as a discipline in early 20th century, the development of management thought dates back to the days when people started working in 'groups' to achieve a common goal. Thus, management is as old as human society. However, it is in the recent years, at the beginning of the last century, that modern management theories were found after drawing the knowledge from various disciplines and by developing its own techniques. During the last few decades the concept of management has changed, especially after industrial revolution, when compelled by the crisis and competition, management was viewed as the key to better and more certain way to succeed. It has been progressively developed throughout these periods of trial and has taken a solid shape through the following seven schools of thought.

Scientific Management

Taylor was considered as the founder of scientific management at the turn of last century. He also gave four principles of management, which will be discussed later in this chapter. Gilberth also contributed to scientific management through motion study as Taylor did through time study. Later Morris Cooke wrote about the applicability of the principles and techniques of scientific management to non-business institutions like municipalities and universities contributing significantly to this school of thought.

Management Theory

Henri Fayol was the first management theorist who was concerned with the principles of organisations and the functions of management. Fayol built up management as a separate body of knowledge, which had universal application to all forms of group activity. He gave fourteen principles of management based on his experiences, which will also be discussed later in this chapter.

Human Relations

This school is based on the contributions of Elton Mayo and his associates, who studied group behaviour and group dynamics. They concluded that feelings or sentiments mould human behaviour and the group exercises a strong influence on individual behaviour and outputs. Financial incentive is less powerful than group standards and personal sentiments in determining the output. They established that human beings are the most important influence for securing higher productivity.

Decision Science School

In the 1950s, the contributions of economists, mathematicians and statisticians or economists as they are jointly called, led to the development of another school of management thought which is known as decision science school. It is concerned with rational decision making by way of defining problems, developing alternatives, evaluating such alternatives and choosing the best solution thereof. According to this school, decision making is the sole way in which managers can discharge their responsibilities of managing and therefore, decision making should be taken as the central focus of the management study. Various mathematical models and analysis like linear programming, critical path scheduling, inventory models, information models, simulation etc. have been developed for quantitative measurement of probable decision outcomes with the help of computers. It is, therefore, not surprising that decision - making has evolved as the key factor in management and is known as the crux of management.

Behavioural Sciences School

Along with the growth of decision science school, there arose the behavioural science school after the World War II. Reversing the findings of the human-relations school, which viewed that happy workers were productive workers, the behavioural-sciences school laid emphasis on human behaviour as it relates to goal achievement and efficiency improvement at work. Keeping this in view, the behavioural approach is mainly concerned with the areas relating to human behaviour, such as, organisational change, motivation and leadership. This school contributed to our knowledge of organisational behaviour by showing the way of integrating individual goals with those of the organisation and managing inter personal conflicts.

Systems Theory

A system is defined as a set of inter-dependent components or subsystems regularly interacting with each other in the organisational environment that constitute the whole organisation. The components are thus connected together and provide the means by which the components interact with one another. Systems theory reconciles the divergent views and approaches, adopted from different schools of thought, integrating the past with present management contributions and explains the complexity and dynamicity of present day organisations. The application of system's approach to the management of home will be discussed later in detail.

Contingency Theory

Contingency theory, also defined as the situational approach, accepts that an organisation affects and is affected by its environment, composed of persons, material and economic resources, market conditions, climate, culture, attitude and laws enacted for their controlled use. The environment thus exercises a potent influence on the enterprise and presents opportunities as well as constraints for the organisation. Successful organisation recognises its important environmental influences and tries to adapt its resources to meet the challenges for achieving the set goals. As a result, effective managerial principles and practices vary with the environment in which the organisation operates, and any generalisations of principles are not practical without taking the environment into account.

Current management thinking is greatly influenced by both the systems and the contingency approaches, which are recognised as the key to effective management. Both these approaches accept the dynamics and complexities of the organisation structures and of the behaviour of their members. However, the management of home is influenced more by system's approach than any other. Therefore, the study and the application of systems approach is relevant in the context of home management too.

SYSTEMS APPROACH TO HOME MANAGEMENT

We have already seen that a system is a set of inter-dependent subsystems. The management of a family also involves the interaction of various subsystems. For this reason we can divide the family management system into three major subsystems, which regularly interact with each other. These subsystems are the three environments surrounding the family as can be seen from figure 3.1. The first one is the household environment, second, the near environment and the third is the larger environment. The management of a family depends upon the components of all these three environments around it.

The **household environment** is the one in which, all the family members interact with each other. It is the most immediate environment, surrounding the family and thus, the most intimate one. The extent of family members' interaction with each other defines the boundaries of this environment. This environment includes human resources of family members like their time, energy, talents, skill, knowledge, attitudes etc. and material resources of the family like house, residential space, furniture, furnishings, other assets and money income of the family members. The family has maximum control over all of their resources of household environment and their management, but the other environments also influence how a family uses these resources. The family identifies its uniqueness with it's household environment, which defines the work and managerial practices, values, goals and standards of each family.

The **near environment** is the environment beyond the household environment, in which a family interacts with other people and groups for educational, vocational, financial, business, religious, political and medical reasons. The near environment of a particular family is dependent upon groups with whom the family members interact, thus, is unique for that family and different for other families. For example, if one family has three members, a husband, a wife and a daughter, their near environment will include the office of the husband, the bank where he deposits his salary, the market from where they shop, the school of their daughter, the religious group of the wife etc. The other family also having three members, a husband, a wife and a daughter may not have a same near environment as the previous one. The reason is that, husbands in both cases might be working in different offices, their daughters might be attending different school, or they might be shopping from different markets, members of different religious groups or they might be pursuing different hobbies. Even if one of this is true, both families will have a different near environment.

The geographical area where the family is located also influences the near environment. This is so because the geographical area determines the community resources available for the use of the family, for example, the type of roads, public transport system, medical, educational and recreational facilities available, and the availability of water and electricity, employment opportunities etc. Thus, near environment also includes the community resources for the use of the family, therefore, influencing the management of the family. When a family shifts its residence to a place where the near environment provides better employment opportunities for the homemaker, the family income will go up but then the homemaker will have less time and energy for household work. This change in the resource availability will also change the home management practices of the family and the whole family will have to adjust to this change in near environment i.e. their locality.

The near environment is less intimate than the household environment and the family has lesser control over it, therefore, they would rather need to adjust to their near environment. But at the same time unlike household environment, the families can choose and change their near environment by shifting their residence to another geographical area.

The **larger environment** extends onwards from the near environment. The boundaries of this environment include all those institutions, which directly or indirectly affect the family. Thus, larger environment of the family may include the state, region, nation or even the whole world depending upon the interactions of the family. For example, for a tribal family living in isolation, the larger environment could be restricted to their own clan or tribe or perhaps to the tribal community. For another person living in a city, affected by the government policies and programmes, the larger environment will extend to the whole nation. On the other hand, a person working in a multinational company and travelling to different parts of the globe will have a larger environment including all those countries where he travels as well. Families are usually less aware of their larger environment but at the same time the cultural practices, norms, ethos and taboos of the family experience the influences of the larger environment on it.

Thus, a system's approach explains that the management of a family as a unit is influenced by other subsystems or environments and their interaction with each other and the family.

Fig. 3.1: Family and its Surrounding Environments.

MANAGEMENT PRINCIPLES

Management is today all pervasive. In fact, it is required not only in family, business, industrial and commercial activities, but also in all human behaviour and activities. Principles of management are applied for accomplishing higher output at home or a factory, a club, a political party or national planning. Thus, from efficient home management to national growth in any country, all management activities to a large extent are dependent on the application of sound principles and practices.

Many scholars and writers from several disciplines have adopted divergent approaches to formulate management principles. Let us now look at the contributions of some of these eminent people whose principles are relevant and universally applicable to all organisations even today. In this chapter we will be dealing with the contributions and principles of Taylor and Fayol and will see how they can be applied in home management.

Contributions of Frederick W. Talyor (1856-1915)

In the latter part of the 19th century, Frederick W. Talyor contributed enormously to the development of management thought. He laid the foundation of the scientific management school and so he is called the **Father of Scientific Management**. Taylor's major concern was that of increasing efficiency in production so that costs can be lowered and profits can be increased while it is still possible to increase the pay for workers through higher productivity. He also believed that the applications of scientific method could yield higher productivity. Keeping these in mind, he gave mainly four principles of management, as discussed below:

Taylor's Principles of Management: The fundamental principles underlying the scientific management as stated by Taylor may be summarized as:

- The development of the best work method.
- The selection and development of workers.
- The relating and bringing together of the best work method and the selection and trained workers.
- The close cooperation of managers and non-managers including clear-cut-division of work and the managers' responsibility for planning the work.

From the above four principles as pointed out by Taylor, it becomes clear that managers and non-managers must have complete understanding about the quality and the quantity of the work to be accomplished. Thus in Taylor's work, there runs a strong humanistic theme. He had an idealistic notion that the interests of workers, managers and owners should be harmonized. Also Taylor concentrated more on productivity. He stressed on time and motion study and other techniques for measuring work so that higher productivity can be achieved. Taylor's principles can be applied in home management for enhancing the capacities of the homemaker and worker (they are usually the same person in the home) and work method for better achievement of family goals by spending least amount of family resources, both human and non human. In home management these principles can be applied as

- The development of the best work methods for doing household work using work organisation and simplification techniques.
- The development of the homemaker for enhancing their quality as human resources and for realising potentials of other family members and converting them into the family's important human resource.
- The relating and bringing together of the best work method for the homemaker and the train them into efficient home-managers.
- The close cooperation of homemaker, other family members and hired domestic help to enable clear-cut-division of responsibility for planning (by the homemaker) and execution of the work (by all of them) among the family members.

Contributions of Henri Fayol

Henri Fayol is claimed to be the **father of modern management**. He also made a valuable contribution to management thought and practice. Unlike Taylor, he considered an enterprise as a whole and did not devote his attention simply upon any single part of it. He evolved what came to be called as *the concept of management as a component of functions.*

Fayol also laid down the principles of management which are based on his experience and keeping in mind, that the principles of management are flexible, not absolute and must be, usable regardless of changing and special conditions. Thus he felt that these basic principles would be used in all management situations irrespective of the organisational framework.

Fayol's Principles of Management and their Application to Home Management: Fayol listed fourteen basic principles of management, which are as follows:

- **Division of Work:** This is very important for using the labour efficiently. Fayol stresses that division of work results in better work and more production with same efforts. He applies this principal to all kinds of work-managerial as well as technical.

  In home management this principle helps a homemaker in dividing the work among every family member and hired domestic help according to their capacities, skills and time availability.

- **Authority and Responsibility:** He explains authority as the right to give orders. Fayol considers authority and responsibility as correlated resources. Responsibility arises out of authority and is also a part of it. Here Fayol also considers authority as a *combination of official and the personal factors*, which are compounded of intelligence, experience, moral worth, past service, etc.

In the management of home, usually authority and responsibilities lies with the homemaker. However, when she is delegating responsibility of any task to any other member, she should also give necessary authority to the performer to accomplish it. For example, if a daughter has been delegated a task of buying groceries, she should have authority to use the money allocated for the selection and purchase of these items.

- **Discipline:** He sees discipline as respect for agreements, which are directed at achieving obedience, application, energy and the outward marks of respect.

  In home management self-discipline of homemaker and every other family member, including that of a hired domestic help is a necessity for the smooth running of the family.

- **Unity of Command:** This means that employees should receive orders from one superior only. He feels that dual command is a source of conflict. This will result in non-accomplishment of work and uneasiness among the authorities as well as the workers.

  At home, a home manager alone should give direction to others, however, this role of home manager may keep shifting from one person to another. For example, a wife could be managing the purchase of daily groceries but for purchasing consumer durable, the managerial authority may rest with the husband or for planning dresses for forthcoming wedding, it may be a grownup daughter's responsibility.

- **Unity of Direction:** According to this principle, each group of activities with the same objective must have one head and one plan. This means that one type of plan having the same objectives should be entrusted with one person.

  Though at home the hierarchy like other formal organisations is not present, but even then only one person should give directions for one type of work. For example, if a paid help is hired to clean the house, one person should direct her. If both husband and wife will start giving directions, thus they both will spend not only energy and time, but also add to the conflict of the hired domestic help. Thus, only one person should guide the hired help in performing the work.

- **Subordination of Individual Interest to General Interest:** The interest of the company should be more important than that of the individual when the two are found to differ, management must reconcile them.

  The family goals are more important in the family living than the personal interest of any one member, and if there is a conflict between them, then it is for a homemaker to resolve them and find ways so that personal interest can be accomplished along with that of family. For example, if a husband wants to indulge himself with a hobby rather than helping homemaker in caring for an infant, it is for a homemaker to see that he does that without neglecting the child.

- **Remuneration:** Remuneration is the price for services rendered by employees. Remuneration and methods of payment should be fair and offer the maximum possible satisfaction to both the employees and the employer.

No remuneration is offered to homemaker for her services as home is not a business organisation and homemaker is not hired to do the job. However, if domestic help is engaged to help a homemaker to do household work an appropriate remuneration should be given by the homemaker. In the case of a homemaker, the remuneration may be her own satisfaction and the recognition she receives from the others.

- **Centralisation:** Fayol refers this to the extent to which authority is concentrated or dispensed with. Individual circumstances will determine the degree that will give the best overall yield. Fayol says that a certain degree of centralisation is found in each organisation, but to yield the results in optimum degree, it should be there.

  The authority at home is also usually centralised with the head of the family and the homemaker. However, in case the homemaker, if also a head of the family, authority is concentrated to one person. Authority can be dispersed at home by delegating responsibilities to other members of the family and assigning necessary authority with it. However, the authority to plan should rest with one person only.

- **Scalar Chain:** Fayol thinks of this as a *chain of superiors*, from the highest to the lowest ranks. All communication should pass through this chain and this chain is lengthy in large concerns.

  At home the scalar chain is small and as there in no hierarchy among family members the communication is more personal than scalar chain dependent.

- **Order:** Fayol follows the simple formula of 'a place for everything (everyone), and everything (everyone) in its (his or her) place'. This is essentially a principle of organisation in the arrangement of things and people.

  This principle is equally relevant at home, where everyone or everything should have a proper place to ensure order. Organisation is important ingredient of home management.

- **Equity to Gain:** Kindliness and justice should be there on the part of the manager to gain loyalty and devotion from the personnel.

  Homemaker should also be equally just and kind towards all members of the family. As a manager, they should not favour any one individual in the family.

- **Stability of Tenure:** Fayol points out the dangers and costs of unnecessary turn over which is the cause and effect of bad management.

  This principle is not directly applicable to home management, however, a secure home base is of most importance for every family member. Especially in times when the rate of divorce is going up and families are disintegrating, stability of the family as a unit is very important.

- **Initiative:** Initiative is the power of thinking out and executing a plan. Since it is one of the 'keenest satisfactions for an intelligent man to experience', Fayol points out that managers should 'sacrifice personal vanity' in order to pursuit the subordinate to exercise it.

Homemaker must provide opportunities to every family member to exercise their initiative. This strengthens the family as a unit and creates a happy environment in the house leading to a better managed home with opportunities for personal growth.

- **Esprit De Corps:** This is the principle that 'in union, there is strength', as well as an extension of the principle of unity of command. He emphasises the need for team work and the importance of communication in obtaining it.

  Building team and establishing communication among the family members is important for attaining the total success of family goals.

Thus from the above discussion it is clear that these principles should be used in all management situations and are very much relevant in the management of home as well.

THE MANAGEMENT PROCESS

Management process helps all homemakers regardless of their particular aptitude or skills engaged in certain interrelated activities in order to achieve their desired goals. As a process it can be easily described as a series of separate parts or five major functions, viz. planning, organisation, leading, controlling and evaluation. According to George R. Terry, *management is a distinct process consisting of planning, organising, leading, evaluating and controlling, performed to determined and accomplish the objectives by the use of people and resources.* These functions or activities are closely interlinked and interwoven and are basic to all managerial activities of the home. All homemakers have to perform these functions. Homemakers use these elements for applications of management concepts to homemaking by planning, organising, leading, controlling and evaluating their work. Let us see this through the diagrammatic sketch given below (Fig. 3.2) that shows the complete management process consisting of five activities that is, planning, organising, leading, controlling and evaluating.

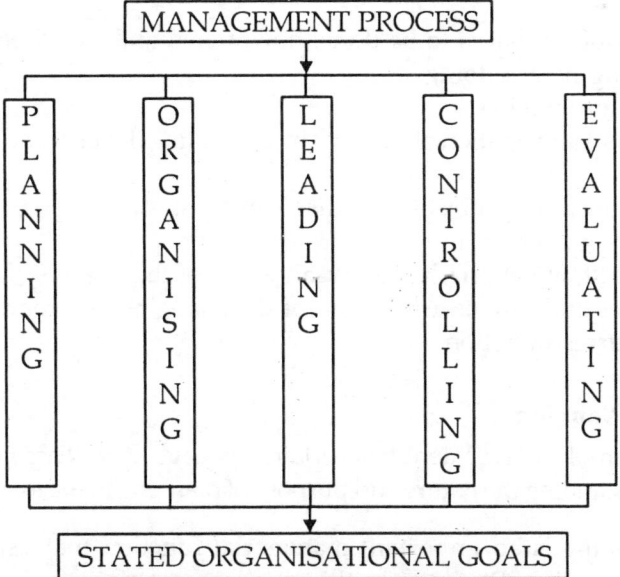

Fig. 3.2: Elements in the Management Process

Planning

Planning is the first and foremost function of management, which needs utmost care and hard work. In keeping with the old saying **well begun is half done**, planning as the first step should be done well. For this, a homemaker will need to think about goals and actions in advance. But the future is uncertain, therefore, she must make proper predictions and then prepare the plans to avoid these uncertainties. The homemaker can make these plans with the help of the planning process, which is logical and involves successive steps. First let us see what is planning and how it can be defined.

Le Briton who defined planning as *a predetermined course of action* gave the simplest definition of planning. He emphasised upon the primacy and forecasting in planning so that the future course of action can be determined in advance. Similarly, Terry said that *planning is selecting and relating of facts and making and using assumptions regarding the future in visualisation and formulation of proposed activity believed necessary to achieve desired result*. However, Maloch and Deacon emphasised the role of decision making in planning by defining it *as a series of decisions concerning standards and/or a sequence of action*.

Gross and Crandall applied these definitions to home management and defined planning as *specifying how family goals are to be reached*. Thus, planning is a process by which a homemaker anticipates the future and discovers alternative courses of action open to her. She then consciously determines the future course of action to achieve the desired results by choosing a course of action from all available alternatives with the greatest economy and certainty. Therefore, by planning a homemaker sets off uncertainty and chance.

Thus, planning is the basic management function, which sets goals as the guidelines for actions. It is the rational and orderly thinking of ways and means for the realisation of predetermined goals. It focuses attention on objectives to provide coordinated systematic map for future activities. It cuts upon the losses due to uncertainty and makes achievement of goals more economical and efficient. Nirmal Singh has summed up some key elements of planning. They are

- Advance decision—what is to be done, how, when and by whom
- Logical thinking before doing
- Involves decision making
- Visualisation and anticipation of future—by evaluating the past and assessing the present
- Futurism
- Process of determining objectives, discovering alternatives courses of action and selecting the best from them
- Process of identifying strengths and weaknesses of the organisation and relating them to opportunities and threats arising out of the changing environment and discovering alternative courses of action.

Characteristics of Planning

Planning as a function of management has certain characteristics. These characteristics help a homemaker in understanding the nature and purpose of planning in the management process.

- **Primacy:** Planning is an important managerial function that usually precedes other functions. Thus it is a first step in the management process. However, it cannot be isolated from other managerial functions, which also have some impact on it.

- **Contribution to Purpose and Objectives:** Planning contributes to the purposes and objectives of the management process. In planning, objectives are clearly stated by braking the goal into smaller achievable targets.

- **Intellectual Activity:** Planning is a mental activity. Rest of the management process involves its execution. Therefore, everything, which is needed to be done, is decided at the planning stage. Thus, it is not an action but an intellectual activity.

- **Continuity:** Planning is a continuous and never ending activity of a homemaker. When one goal is achieved, planning for the next goal starts, e.g. the first goal of the homemaker in the morning is to send children to school. After that, she starts planning for breakfast and then for lunch. Thus, planning is a continuous process. Even during the process of controlling some planning considerations are done to suit the changing situations and also during emergencies, when it calls for a modification in the original plan. Thus planning demands continuity.

- **Flexibility:** Planning leads to the adoption of a specific course of action and the rejection of other possibilities. When the future cannot be moulded to conform to the course of action, the flexibility is to be imagined in planning by way of adopting the course of action according to the demands of current situations.

- **Unity:** Every family member makes his/her individual plans. Maintenance of consistency or unity of everyone's plans is an essential requirement of the planning of a homemaker. Therefore, whenever the homemaker plans, she should consider the individual plans of other family members and generate their consent to maintain unity in planning as well as in executing.

- **Precision:** Planning must be as precise as to its meaning, scope and nature. It should be framed in intelligible and meaningful terms by way of specifying the expected results. For example, the budget prepared by the homemaker for the family should be precise as far as possible so as to reach their goals conveniently.

- **Pervasiveness:** Planning is a pervasive activity covering the entire family and all aspect of family living. It is not the exclusive responsibility of homemaker alone. In the family each and every member should be directly involved while planning as it affects the future of every member and that of a family as a unit.

Process of Planning

The process of planning can be better understood with the help of successive steps. By following these steps (figure 3.3), which can be easily executed, a homemaker can make a comprehensive plan. Now let us take an example of step by step planning for savings money for higher education of the children.

1. **Being Aware of the Opportunities:** Although it precedes actual planning and is therefore not strictly a part of the planning process, an awareness of opportunities in the household, near or larger environment is the real starting point for planning. It involves the assessment of strength and weaknesses, understand what problems we wish to

solve and why, and know what we wish to gain. Planning requires realistic diagnosis of the opportunities and situation. Thus, before planning to save money, a homemaker should be aware of all of her resources (human, non-human and community resources), e.g. an ability to take up a job herself and other opportunities in her near environment for generating more resources like transport facility etc.

2. **Determination of Objectives:** The next step is to establish the overall objectives or goals of the family, for long term as well as short term. Objectives specify the expected results and determine the pattern of the proposed course of action. The major family goals are broken down in smaller objectives and they are further divided into series of tasks. Determining objectives indicate what is needed to be done. Thus, while making a saving plan, a homemaker should clearly state how much money she will need and when. This will help her in determining the amount of money she must save every month.

3. **Developing Planning Premises:** Planning premises are assumptions about the future environment in which the plan is expected to be executed. They explain the settings in which all the activities will be carried out. Planning premises provide important facts and information for forecasting the future trends and basis for analysis of the situation. A homemaker planning to save money for children's higher education will need to premise regarding the possible educational courses her children may like to do later on and approximate money that would be needed for it and therefore how much money she must save every month, and by then.

4. **Determining Alternative Courses of Action:** The next step is searching and developing various courses of action. This also means carefully examining and selecting all the possible courses of action, especially those, which are not immediately apparent. For example, a homemaker will need to decide how she will save so much money every month. The possible courses could be that she can reduce food expenditure, travel expenditure and entertainment expenditure or increase her income by doing part time or full time job, depending upon her skills, opportunities and requirements.

5. **Evaluating Alternative Courses of Action:** Now each alternative is examined for its strengths and weaknesses. They are evaluated by weighing them in the light of premises and goals. A homemaker should now carefully evaluate all alternatives in the light of resources available. For example, if she reduces expenditure on food, clothing, travel and entertainment then she may not be able to afford the life style she wishes to have. On the other hand, if she takes up a job then all the family members will need to adjust accordingly and she will have to delegate more household work to them.

6. **Selecting a Course of Action:** The best suitable alternative is selected by making a decision. Thus, at this stage important decisions are taken regarding the future course of action. These decisions help in forming an operative plan, which will be executed. The homemaker has to decide now whether she will cut on daily expenditure or increase her income by taking up a job. Suppose she decides to increase her income so that the family can afford desired lifestyle and simultaneously save money for the higher education of the children, then she has selected a course of action for achieving her goal.

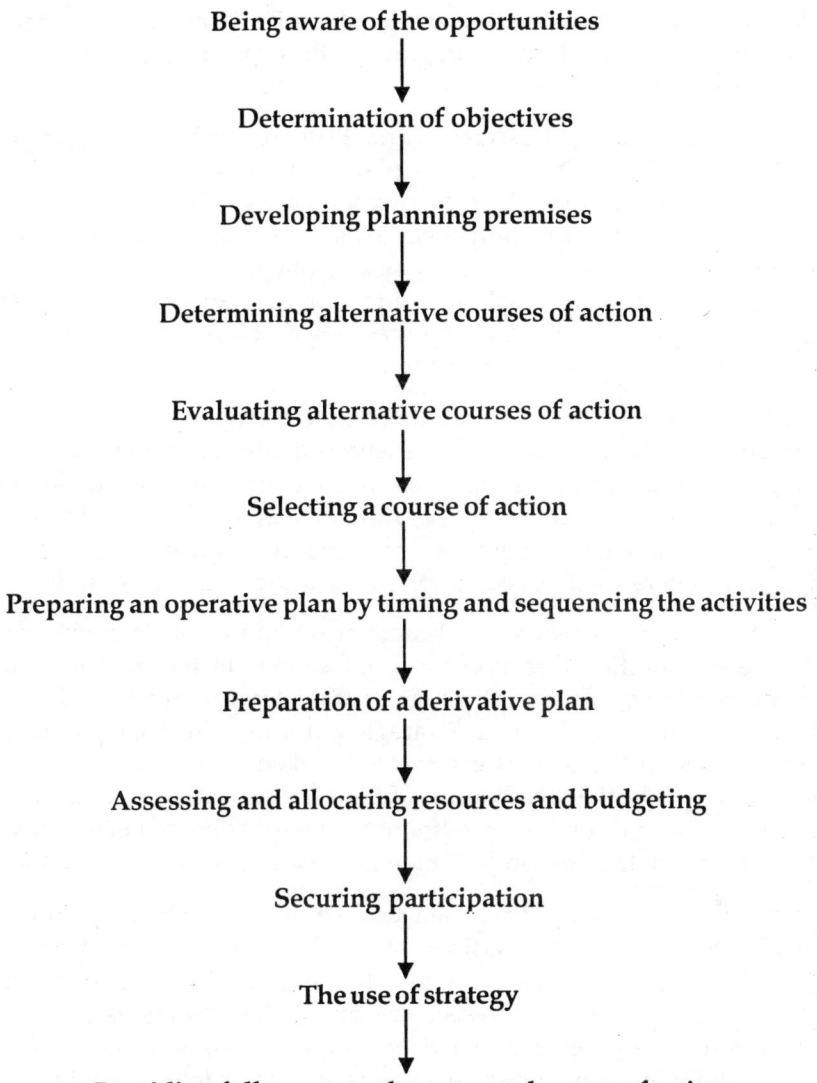

Being aware of the opportunities

↓

Determination of objectives

↓

Developing planning premises

↓

Determining alternative courses of action

↓

Evaluating alternative courses of action

↓

Selecting a course of action

↓

Preparing an operative plan by timing and sequencing the activities

↓

Preparation of a derivative plan

↓

Assessing and allocating resources and budgeting

↓

Securing participation

↓

The use of strategy

↓

Providing follow ups to the proposed course of action

Fig. 3.3: Steps involved in Planning Process

7. **Preparing an Operative Plan by Timing and Sequencing the Activities:** The selected courses are sequenced and timed to prepare an operative plan. The starting and the finishing time is fixed for each activity to maintain the smooth flow of work. For example, homemaker will now prepare a plan by listing all the activities she need to carry out like searching for job, delegating household work among other family members etc.

8. **Preparation of a Derivative Plan:** These are the supportive plans, which are prepared for each activity. They are developed to integrate operative plan with each area or segment of the organisation, its relevant policies, procedures and programmes. For example, a homemaker will now prepare derivative plans for leaving on time for

work, doing household work with the help of family members and/or paid help, care of children in her absence etc. and integrate all these plans for the smooth and problem free management of family.

9. **Assessing and Allocating Resources and Budgeting:** Budgeting or assessment of necessary resources for carrying out a plan is the next step. To achieve any target some resources are required. How these resources will be procured, where will they come from, the flow of these resources etc. are some questions which are now answered. Therefore, now she will decide the amount of money she would be spending on hiring paid help, buying clothing for self, travel to work etc. and also, the amount of money she must earn to enable her in meeting these expenses and saving the desired amount of money for higher education of children.

10. **Securing Participation:** The plans are more effectively executed if the participation of other members of the organisation is secured at the planning stage. This involves communication of important details of the plans to increase their understanding of the expected results. Participation improves the quality of planning by additional view points. Therefore, now the homemaker must communicate her plan to the members of the family and secure their participation in planning and executing it.

11. **The Use of Strategy:** Strategy is a broad programme for achieving organisational goals. It creates a unified direction for organisations in terms of its many objectives and guides resources to move the organisation towards those objectives. Actively formulating a strategy is known as **Strategic Planning**. Strategic planning usually has broad and long term focus. This can be undertaken by the homemaker for planning long term family goals like saving money for higher education and marriage of children, and to ensure regular income after retirement. By applying strategy to her plan she can guide her income, skills, time and attitude etc. towards saving money for future needs.

12. **Providing Follow ups to the Proposed Course of Action:** Since planning is done for future and future cannot be predicted with absolute accuracy, a follow up plan is proposed which is executed only when the situation demands the adjustment of operative plan due to some unforeseen reasons. In the fast changing world, follow up plans are absolute necessity especially for achieving long term goals. This means that a homemaker must review her achievements at regular intervals and if due to certain reasons she could not adhere to her operative plan, then she should adopt a follow up plan. For example, if she could now save enough money even by taking up a job then she can cut on the expenditure on food, clothing, entertainment, travel, paid help etc. as a follow up plan to meet her goals in time.

Role of Planning

Careful planning needs time and skill of a homemaker, therefore, a homemaker must be aware of its benefits. Let us now discuss the reasons for spending resources on planning function.

- **Effective and Faster Goal Achievement:** Planning ensures orderly and meaningful action. The best course of action is selected to reach goals, therefore, resources and time is not wasted in unnecessary activities. Planning guides a family through the definite, safe and shortest way towards its goals.

- **Provides Competitive Edge:** Planning involves premising and forecasting. Thus, it gives a family an edge in the uncertain and rapidly changing situations by selecting the best course of action and orderly progress. Planning makes provisions as it foresees the future, adds strength and builds confidence of the family in meeting its goals.

- **Focus on Objectives:** All the activities in planning are focused towards the achievement of objectives. Thus, it leads to unity of purpose and direction. The activities are coordinated, and isolation, overlapping, duplication and cross purpose working is avoided. Therefore, every effort, resource and action is directed towards the achievement of goals.

- **Minimizes Cost of Performance:** As we have already discussed that planning avoids the overlapping, duplication and cross purpose, the cost of performance due to wasteful activities is reduced. It removes hesitancy, avoids crisis decisions, eliminate false steps and protect against improper deviations. At the same time, it ensures the selection of most profitable course of action. Thus, the performance cost is minimized.

- **Ensures Effectiveness of other Managerial Functions:** Planning provides direction to all other managerial functions. Thus, the execution of plan ensures good organisation, coordination and an effective control.

- **Provides Basis for Control:** A control is always built in the plan. The plan clearly states the expected results and the deviation from them shows that the activities need to be controlled from time to time. Thus, control confirms that a plan is correctly followed and the targets are achieved in time.

Limitations of Planning

In spite of so many benefits, planning has few limitations as well. Being aware of these limitations can help a homemaker in overcoming them.

- **Accuracy Depends on Premises:** The planning is for future and future is neither certain nor completely predictable. Thus, planning relies heavily on the premises based on the information and data. If dependable data is not available for premising, the planning cannot be accurate. As the duration of planning increases the accuracy of planning decreases. For example a child might want to go for different course for higher studies than planned by the parents, the cost of the desired courses could actually increase much more than anticipated by the family, or the government policy like that of privatization of education can make higher education to demand more money than saved by the family.

- **Leads to Inflexibility:** Planning limits freedom by predetermining the course of action. Thus, it leads to rigidity, curbs initiative, creativity and freedom. The scope of personal development is also reduced. People are bounded by procedures, policies and programmes with little space for individuality and self-expression. For example, while following a plan to save money for future, a homemaker has left little space for flexible working hours or choice of method of doing household work.

- **Difficult in Fast Changing Scenario:** In the ever changing environment, planning has become difficult especially for the long term. People prefer to work on day-to-day basis, which is faster and more economical. For example, a homemaker can never be certain of the availability of educational opportunities for her children when they will grow up and the money they would be needing for enabling the child in selecting the course of their choice.

- **Time Consuming and Expensive:** Though a good plan is always beneficial, but many people spend more time, money and energy in planning rather than its execution, making it expensive and time consuming. This is more of a limitation of the manager than that of the planning function itself.

    In spite of these limitations planning is the most important function of management, which helps in reaching goals with the help of decision making. It involves selecting the courses of action that will follow. It is planning which answers the questions like 'what to do', 'how to do it', 'when to do it' and 'who is to do it' in advance and thus, bridges the gap between the starting point and the target one wants to achieve.

Organising

For effective performance, there is a great need for

- developing an intentional structure of roles of the family members, and
- securing the co-ordination of individual efforts towards achieving common family goals.

The second function of the management, i.e. organising aims at achieving these two factors. Similar to the other functions of management, there is no single best way for organising. There is a general need to understand the organisation structure and to put the principles to practice. One has to think of ways and means that would work and give best results under such circumstances.

As said earlier, after planning, organising is the second important function of management. It helps in clearly stating the duties of each family member, thus reducing vagueness and brings clarity. Once a homemaker has established objectives, and developed plans to reach family goals, she must design and develop an organisation that will enable her to execute her plans successfully.

William F. Glueck states that *organisation is the process in which people and the tasks they perform are related to each other systematically to achieve the enterprise objectives.* Organisation includes dividing up the work among groups and individuals (division of labour) and linking the sub parts together (co-ordination). Similarly Oliver Shelden points out that *organisation is the process of combining the work which individuals or groups have to perform with the facilities necessary for its execution, that the duties so performed provides the best channels for the efficient, systematic, faster co-ordination and application of the available efforts.*

Kimball explains that *organisation is subsidiary to management. It embraces the duties of designing their functions and specifying the relations that are to exist between departments and individuals.* Davis, while emphasizing the significance of relationships in an organisation and the influence of planning upon it goes on to state that *organisation is the function of creating and providing the basic conditions and relationships that are a pre-requisite for effective economic execution of the plan.* Consequently an organisation includes an advance provisioning and propositioning of basic factors and potential factors, as specified by the plan.

While illustrating the view point of relationships and their structural setting, Spriengel states that in *its broadest sense, organisation refers to the relationship between the various factors present in given endeavours.* From the standpoint of the enterprise as a whole, organisation is the structural relationship between the various factors in an enterprise. Nirmal Singh sums up by stating that *organisation is the setting up of functions to be performed by individuals or groups respectively, for the attainment of the defined objective given the necessary authority and responsibility.*

Gross and Crandell explains *organisation as an orderly design a homemaker creates by planning and coordinating the activities of the home.* They further stress that a homemaker has to organise the activities for self and for other as well.

Thus, organising involves allocating the tasks to different individuals who in turn make efforts collectively in order to reach to a common goal. For example, a homemaker may decide the course of actions, but it is actually to be implemented by her family members. To make organisation successful, the homemaker must divide the activities into units and each unit is allotted to the person depending on his interest, experience, skill, capabilities etc. It is necessary that all people should aim at achievement of family objectives and the individual interest should not be over group interest. The homemaker should make comprehensive units having the same type of work and should choose the right person for the right job. For example, shopping for groceries and vegetables can be clubbed and assigned to one person who is responsible for meal planning. Thus, organisation helps to increase the efficiency and leads to more specialisation and increases the output. Due emphasis can be placed on each unit. It helps to inculcate team-work and provide a scope for developing managerial abilities. It also helps to increase promptness and certainty in work. For this it involves work simplification, sequencing or pattern, managing efforts of others — all behaviour of all the participants, formalise internal structure of roles and positions.

Organising involves establishing an intentional structure of roles for people in any enterprise with a hope that all the tasks necessary to accomplish goals are assigned to people who can do them in the best possible way. For example, in a home, after making the budget, the housewife might allocate the responsibility of purchasing furniture to the husband who would be a better buyer, while purchasing clothing by herself as she can do it better than her husband.

Organising also means that a homemaker coordinates the human and material resources of the family. It necessitates that she must organise family members, materials, tasks and time. The effectiveness of an organisation depends on the ability of the homemaker to use family resources to attain its goals. Thus, during this process, proper relationship among work, people and other resources are established and the authority and responsibility are channelised. Therefore organising is an inseparable part of managerial action.

Characteristics of Organising as a Function

Efficient organising will contribute to the success of an enterprise and will have the following characteristics:

- It embodies the goals of the family.
- It helps to show the breakdown of goals.
- Helps each family member to see his tasks and resources he has to accomplish it.
- Tells each family member where his accountability lies and who are in his sphere of command.
- Family members know what communication to others is demanded of him and vise versa.

- Family members become aware of their rights and responsibilities.
- Helps vigilance against loss of resources and mis-direction, and reinforce economy and effort.
- Helps to secure the best out of the team of family members and adds the essential flexibility.
- Ensures work efficiency and order.
- Has human orientation rather than work orientation.

Organising Process

Organising as a process involves certain steps. For example, when a homemaker has decided to take up a job to increase family income so that money can be saved for the higher education of the children as planned by her, she has to organise her household work so that she can work outside home. In that case, she needs to decide on the following steps:

1. **Identifying and Classifying Activities:** The first step is breaking up entire work in different segments. Therefore, a homemaker must list all the activities which need to be done like buying groceries, vegetables etc., cooking meals, cleaning of the house, washing clothes and cleaning utensils.

2. **Grouping Activities:** Now all activities must be grouped in the light of human and material resources available and the best way of using them. Therefore, a homemaker should classify the activities which could be carried by herself in the limited time, by her husband, by each child and the activities for which she need to hire a paid domestic help.

3. **Delegating Work:** Each group of activities can be delegated to a person who can accomplish that work in the best possible manner. Each family member should also be assigned authority and resources necessary to perform these activities. Thus, delegation is making a scheme of work and assigning it to others rather than doing everything by the homemaker herself. Delegation also has certain steps which can clearly explain how it should be done. These steps are:

   - Identifying the total tasks of the homemaker
   - Identify what part of it she can and should perform herself measured by the yardstick of organisational effectiveness
   - Delegate the rest, after breaking it into clear cut tasks easily understood and recognised by the delegate
   - Assign authority to each delegate
   - Enhance responsibility

4. **Delegation of Authority:** This step involves tyeing all the family members together horizontally and vertically, through authority relationships and information flow. For example, a person planning and cooking meals must have authority to decide the menu and necessary ingredients and communicate her needs to the person who will buy them. Whereas, a person buying those ingredients must have the authority to select them while keeping in mind the needs of the person cooking meals.

5. **Coordination:** Coordinating the efforts of all the family members is very important to achieve goals. For example, when a person cooking meal decides to prepare stuffed capsicum, she must communicate this purpose to the person who will be purchasing the raw ingredients so that she can select medium and same sized capsicums. Thus, coordination between the person buying ingredients and the person cooking meal is essential.

Coordination helps to unify activities and parts of a plan into a harmonious and workable whole. It is necessary for the homemaker to see that all units of work are carried out timely so that the whole activity can be unified. It is necessary to coordinate the efforts of all individuals towards achieving common family objectives.

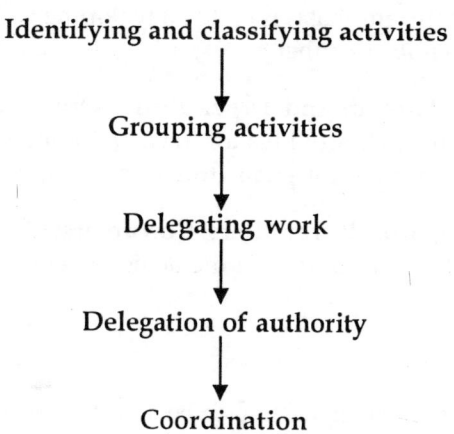

Fig. 3.4: Steps Involved in the Organisation Process

Importance of Organising

Similar to planning, the homemaker should also understand why she must organise. Let us now look into some of the benefits of organising.

- **It Increases Management Efficiency:** Organisation increases the efficiency by avoiding delays and duplications. One task is assigned to only one person and every task is delegated so that no activity is left undone. This ensures efficiency in achieving family goals.

- **Optimum Use of Human Efforts:** Organising utilises the principle of specialisation. The tasks are assigned to the person who is the most capable of doing it. As we have discussed earlier that the housewife might allocate the responsibility of purchasing furniture to the husband who would be a better buyer for it while doing purchase of clothing herself as she can do it better than her husband, thus, making the best use of human effort.

- **Laying Proportionate and Balanced Emphasis on Various Activities:** Organising ensures that due emphasis is given to each activity. Overemphasis on one activity while ignoring or forgetting others should be avoided. For example, while organising, she may delegate the responsibility of dusting the house to the children, but she also has to ensure that they get enough time to study or can rather give some of her time to them to help them in their studies. Thus, she would be balancing between all the activities.

- **Facilitates Coordination:** Organisation helps in coordination of efforts of all the members of the family. It also ensures the smooth flow of information and good communication among all the family members.

- **Provides Scope for Training and Development of Family Members:** Organisation has human focus rather than the work focus. People are given more importance than the actual work. Therefore, emphasis is on the development of skills, attitudes, capacities and creativity of the family members.

- **Helps to Consolidate Growth and Expenditure:** Organisation while regulating the expenditure of family income, provides for the development of all the family members. Therefore, while achieving family goals, the whole family grows together as a unit.

- **Prevent Growth of Laggards, Wirepullers, Corrupters:** Organisation ensures that each member gives his or her contribution in the achievement of family goals, according to one's age and ability.

Staffing

Some management authorities consider staffing to be a part of the organising function. It is the process by which managers select, train, promote and retire subordinates. Thus, it is the recruitment and placement of the qualified personnel needed to do the organisational jobs. Though it is not of much relevance to the homemaker, but she also has to perform this function sometimes when she hires paid help to do domestic chores or a driver to drive the vehicle. It involves the following activities.

- Analysis of the job i.e. job description and man specification
- Selection of the personnel i.e. recruitment, screening and selection
- Training of selected personnel
- Motivation and appraisal
- Placement and promotion
- Leading and supervising

Leading

This is the third function of management. It is the process by which actual performance of family members is guided towards family goals. Leading is a function of human relations and motivations. It can be called by various names such as 'directing', 'supervising', 'motivating', 'guiding', 'executing' or 'actuating'.

The homemaker's responsibility does not end with organising. She has to guide the actual performance of all the family members. Thus, leading describes how homemaker directs and influences others to perform the essential tasks of the family. By establishing the proper atmosphere, she can help everyone to do his/her best. In a family leadership positions could shift among family members thus, no one person dominates others.

Characteristics of Leading

After plans have been made and the organisation structure has been determined, the next step is leading the organisation towards the achievement of its goals. This function of management involves the following activities:

- **Motivating and Guiding Personnel:** Homemaker has to motivate all the family members to work together to reach common family goals. It is for her to see that the morale of the members is high so that work is done quickly and efficiently. She also had to guide each member towards the set target.

- **Influencing and Shaping the Social System:** Family is a smallest unit of society. Work patterns, which are developed at home, also influence the work outside home. Children learn the managerial abilities at home, which they utilise while doing every task throughout their life. Thus, the homemaker also shapes the social system by influencing family members and the family environment.

- **Understanding Followers and Securing their Cooperation:** The leader has to understand the needs and capabilities, strengths and weaknesses, and problems of her followers. This will help her in getting their cooperation in completing their assigned work and contributing in the family goals.

- **Creating Climate for Performance:** A positive and happy environment conducive for working with efficiency should be created by a homemaker. She should build the confidence of her team with appreciation and guidance.

- **Directing Efforts towards Defined Objectives:** A homemaker must influence family members so that they will work willingly and enthusiastically towards the achievement of family goals. To set the family in motion, the team mechanism has to be activated through directing and leading, to carryout the plan of action.

- **To Get Full Co-operation from the Members:** This function of management is very concrete, unlike planning and organisation, as it involves working directly with people. Homemaker should ensure that she gets full cooperation from family members at work. In order to maintain good interpersonal relationship working teams are inspired and motivated to do the work whole heartedly and willingly.

Significant Roles of Leading

Like planning and organising has benefits as functions of management, so does leading. Let us now see why a homemaker must lead.

- **Motivate Group Efforts:** Leaders motivate the group members to make efforts willingly for achieving the family goals. A homemaker should act as leader of her home to ensure that all members reach their targets successfully.

- **Emphasis on Human Performance:** Leading is a function, which emphasises entirely on people. Thus, focus is more on the person doing the work than on the activity. However, leading improves the efficiency of the management process by improving the human performance.

- **Basis for Cooperation:** Leading provides basis for cooperation among the family members. It generates the team spirit and togetherness.

- **Leads to Greater Organisational Effectiveness:** Leading enhances the effectiveness of the organisation by directing all the efforts towards the goals.

- **Efficiency:** Efficiency of work improves when people work willingly with enthusiasm. Leading improves the morale of all the members of the team.

- **Ensures Performance According to Predetermined Goals:** Leader generates cooperation and unifies the efforts of everyone towards set goals. The emphasis is laid on group efforts and achievement of common goals.

- **Forward Looking:** This function is forwards looking as it directs all members towards the achievement in future. Leading does not talk about the past, it helps members see the future i.e. accomplished goals.

Supervising

When the work is delegated to others supervision becomes important. Rarely a person manages a home without the help of family members or paid help. Home managers delegates tasks like shopping to the husband, cleaning the rooms to the children and sweeping and mopping to the servants. After assigning these tasks, homemaker has to supervise the actual performance of work. Depending upon the performer and the situation, the supervision may include giving directions, guiding, explaining, instructing or demonstrating. There are two basic types of supervision, depending upon whether the focus is on work or the worker. They are guiding and directing.

- **Guiding:** When the focus of supervision is on the performer's development, it is called guiding. In guiding the major interest is upon what is happening to the individual carrying out a process. In a situation where a person is learning a method of work as in the case of children helping in doing household work, the focus is on their growth and welfare. Thus the experience of a person are more important.

- **Directing:** When the end product is of most importance emphasis is placed on the process itself and this type of supervision is called directing. Therefore, clear instructions and orders are given to assume the understanding of the method of doing a job. It is especially helpful when speed in achievement or quality of end product is important.

As can be seen, manager as a leader, plays a very important role in an organisation. For all these reasons, the roles of managers are dealt with details separately in the chapter on Managers and Management.

Controlling

This is the fourth function of management. It can be defined as *the process that measures performance of activities and guides it toward the goal that is pre-determined. It involves the implementation of the plan with proper controlling measures.*

Controlling is also defined as a measurement of accomplishment against a standard and correcting any deviation in attainment of objectives. Controlling ensures both qualitative and quantitative performance of work in the family as planned. As soon as a plan is put into action, control becomes necessary. It helps the homemaker in seeing **what is being done is what was intended to be done**. Control tends to measure the progress made in action against a standard, indicates any deviation from planned action and suggests steps to correct deviation by following alternate courses of action. Thus, the principle underlying it is **a stitch in time saves nine**.

Characteristics of Control

- **Leading Function:** Control becomes the leading function of management as it starts planning and continues till the goals are reached. Because of the integrated character of managerial functions, it is rather difficult to isolate control from other functions of management.

- **Unity:** Control is an integral part of management, and is exercised by all managers at all levels. Every member of the family has to exercise control by identifying deviations in time and correcting them so that the targets are achieved.

- **Continuity:** Like other functions of management, control is an ongoing activity on the part of all homemakers. For example, while following the time plan, the homemaker has to check it continuously to see that what she is doing is according to what she has planned.

- **Flexibility:** It is as much an important attribute of controlling as it is for planning. When plans are required to be revised to meet the needs of time and situation, control ensures a change in the plan and is made to be capable of adjusting to the needs of changing time and situation.

- **Pervasiveness:** Like planning control is also a pervasive activity. Control is exercised through out the length and breath of the organisation. At all levels and for all areas of the family living, control has a pervasive use to get things done by others.

- **Forward Looking:** To make the performance conform to plan, control must be forward looking in character. Its focus is not on finding deviation but on corrective measures to ensure performance.

- **People Focused:** The modern system of control has become performer focused rather than work focused in character. It also persues a better type of managing by setting objectives and adopting self control.

Process of Control

Like planning and organising, controlling as a function of management can be done step by step. Let us take the same example of saving money for children's higher education to see how control can help a homemaker in ensuring that the goal is achieved.

1. **Establishing Standards:** It involves laying standards or goals so that performance can be measured. These standards are laid down while planning. These standards are simply the criteria of performance. In the entire plan critical points are selected where the performance is checked against the standard laid down. For example, while making a plan a homemaker must have calculated that the amount of money she should save every month, which will act as her standard. Therefore, she may select the end of each month as a critical point when she will check how much money she has managed to save and compare.

2. **Measuring Performance against Standards:** Measuring performance against the standards help in finding the deviation, if any. Thus, these periodical checking show that efforts are moving in the right direction. A homemaker will compare the amount of money saved by her during the month against set standard. If she finds deviation, i.e. she has not managed to save as much as she should have, then she will go to the next step. If she finds that she has managed to save the amount of money she planned, then she is assured that she would be able to save enough money for her children's higher education. Also, she is assured that her efforts are in the right direction.

3. **Correction of the Deviation:** After step one and two if deviation is found, then the corrective measures must to taken. For adjusting these deviations the manager must take measures to correct them by suggesting alternate course of action. This can be done by redrawing the plan, modifying goals or by re-assigning the duties. For example, if a homemaker finds that she has not managed to save the required amount of money, then she should either make changes in the plan by reducing the expenditure wherever possible, or re-assign the duties of those work where more than the planned money is spent. She can also modify her goals, if she cannot find any other way to reach the target.

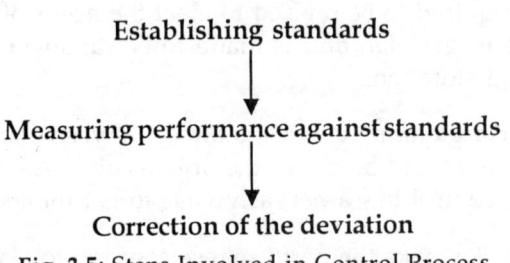

Establishing standards

↓

Measuring performance against standards

↓

Correction of the deviation

Fig. 3.5: Steps Involved in Control Process

Role of Control

Similar to other functions of management, control is also important. A homemaker must know the benefits of controlling.

- **Ensures Performance According to Pre-determined Standards and Goals:** Control checks the performance against predetermined goals, thus, monitors the performance against standards. It ensures by correcting deviations if any, that the goals are reached.

- **Rescues and Provide Ways and Means for the Information Flow:** The role of control is to rescue whenever a problem arises. Thus, it saves the plans from sinking and helps in detecting the problems and provides means for rising out of it.

- **Efficiency:** It improves efficiency by saving time, money and energy. By correcting deviation in time it ensures that goals are reached in spite of changes in situations and problems, and time, energy and effort invested in the plan are not wasted.

- **Aids Coordination among all Activities by Regulating Them:** Coordination is to be established among all activities. Control unifies the efforts and creates harmony among all the members.

- **Forward Looking:** Control is undertaken for correcting deviation to achieve goals. The reason of finding deviation is neither to blame performer nor to fret about the past. It is forward looking and is exercised to rectify the problems and ensure the achievement of goals in future.

- **Leads to Greater Organisational Effectiveness:** Control improves organisational effectiveness by contributing to the achievement of goals. It ensures performance and detects and rescues when problems arise.

Techniques of Control

They are those methods, which help in detecting the deviation in the performance. Primarily there are two types of methods. These are

- Budgetary
- Non budgetary

Budgetary Methods: These methods involve statement of anticipated results in numerical terms. Under this category the most important techniques, which can be used by the homemaker to control her plans are following:

- **Income and Expenditure Budget:** This type of budget shows the family income from various sources and its distribution on family expenditure on various items for a specified period of time. The details of this technique are discussed in the chapter on money management.

- **Cash Budget:** This type of budget shows how much cash is available for expenditure and how much is already being consumed. This budget helps a homemaker in checking the flow of cash.

Non Budgetary Methods: These methods do not involve numeric but help in checking the activities necessary for achieving goals. Some of the most useful such techniques are following:

- **Checklist:** They can also be further divided into two types depending upon how a homemaker uses them.

  - **Mental Checklists:** Most often homemakers prepare a checklist and mentally check, which activities are done and which are left. However, there is always a chance that she may forget one or more vital items of the checklist. But it is used most often and is quick and inexpensive way of checking especially when the activities are routine and not very important.

➢ **Records:** This is a written form of checklist in which jobs, activities and even expenditure is recorded on a sheet of paper. The jobs completed are entered in the record and thus, a homemaker can assess the performance just by looking at a record. This method eliminates the error in checking due to forgetting but needs more time, energy and attitude of homemaker for writing the records.

• **Special Reports and Analysis:** They can be undertaken only when a homemaker wishes to make a special analysis, which is very important for her. For example, she wish to assess the variation in expenditure during the summer vacations when many guests come to stay in her house. Thus, she can prepare a special record of added needs and expenditure during those months so that she can balance her expenditure in time without disturbing the goals.

• **Time Event Network Analysis:** They are specialised techniques which relate time and events. Though they are not used often by homemakers but are very helpful in studying their work methods.

• **Personal Observations:** They are the most frequently used method of checking. However, the objectivity and effectiveness of using this method depends entirely on the person using this technique.

• **Operational Audits and Statistical Data:** Though these methods are not used by homemakers but are important controlling devices. They are used in bigger and highly specialised organisations to check the performance.

Control as a Feedback System

Feedback is a process in which the result of the performance is utilized to adapt the process again so that the goals can be achieved. Figure 3.6 shows the simple feedback.

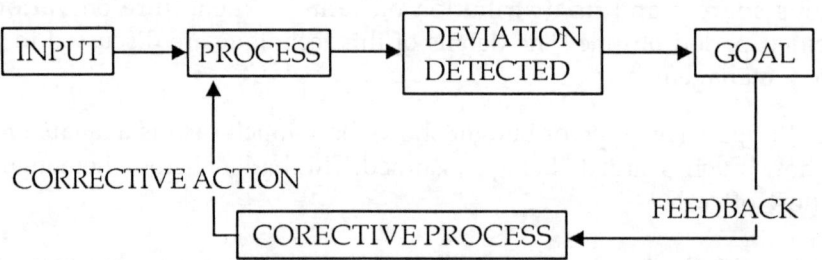

Fig. 3.6: Control as Simple Feedback

The control function enables a manager in measuring the actual performance and compare it against pre determined standards. The deviations are then identified and analysed to suggest corrections. These corrective actions are implemented in the plan in order to achieve the desired results. Thus control acts as feedback. Therefore, feedback measures output of a process and feeds it into the system as the input after correcting deviation to obtain the desired output. For example, a homemaker checks at the end of the month that she has spent more money. She will use it as feedback and correct her expenditure pattern for the next month so that by the end of the next month she is able to save the money she planned to save.

Control as Feedforward

Simple feedback control always cannot provide effective control. For example, if a homemaker finds out at the end of the month that she has over spent then she cannot correct her actions and reach her saving target for that month. Therefore, she needs future directed control i.e. feedforward. However, if she monitors her expenditure and while spending finds out the deviation and corrects it in time she is using control as feedforward. It is recognised that the problems that are best solved are those seen and can exercise effective control only if they can see deviations coming in time. It is therefore wrong to say that planning is looking forward and control is looking backwards. Effective control is also forward looking and is therefore called feedforward. Feedforward monitors the inputs into a process to ascertain whether the inputs are as planned and if they are not, then either inputs or process is changed in order to assure the desired results. Figure 3.7 shows the relationship between feedback and feedforward.

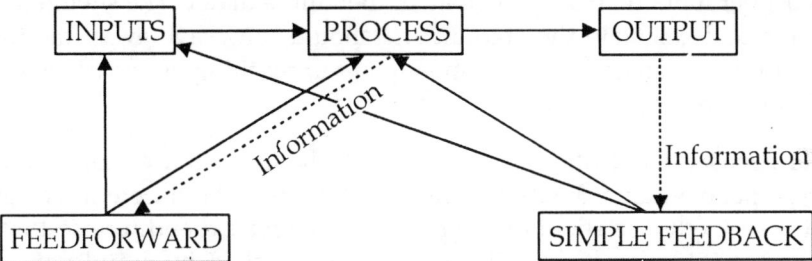

Fig. 3.7: Feedback and Feedforward Control Systems

Requirements for Effective Control System

The objective of control is to find and correct deviation so that goals can be reached. While planning an effective control a homemaker must keep the following points in her mind.

- **Control Should be Tailored into the Plan:** The objectives are set in plans and goals are defined. Control uses these objectives as standards to check for deviation. Therefore, while planning, a homemaker should decide the checking points when the performance can be measured and if any deviation is found, it can be corrected in time. Plans are made for future, on the basis of certain assumptions. However, future is uncertain. Tailoring control in the plan helps in altering the plan in the light of new information when the deviation occurs due to differences in the situation from that premised. For example, a homemaker while planning savings for higher education for children can decide that at the end of each month she must save a certain amount to reach her goals and control her plan by checking it against the actual saving each month.

- **Establish Clear and Objective Standards or Goals:** Control checks deviation against set standards and pre-determined goals. Thus, the control standards should be very clear and objective. A homemaker must state a fixed amount that must be saved every month, which can be very clearly understood and has no ambiguity or bias in it.

- **Control at Critical Points:** Effective control should be made at those points, which are critical for appraising performance. These critical points are turning points as well. If deviation is not corrected at these points then it is not possible to reach goals and the whole exercise of controlling becomes futile. Thus, the points when the deviation if any, can still be corrected should be the checking points for control.

- **Selecting Suitable Checking Devices:** Appropriate checking devices must be used for finding deviation. It is for homemaker to select the most suitable control device, which can be handled easily and effectively. For example, a husband can use complex balance sheet to check his expenditure, which cannot be used by children. A simple checklist will be more suitable control device for them.

- **Promptness in Control:** New decisions should be taken promptly in the light of new information so that courses of action could be altered in time and goals can still be reached. For example, a homemaker finds out the course her child wishes to pursue may cost her more than she anticipated, then immediately she should assess her needs of saving and calculate the amount she should now save every month to reach her goal. Prompt action and control can help her in reaching her goals and coping up with the changes in the situation.

- **People Focused Control:** The control should be tailored according to the person. For different people doing the same work, control could be different. For example, every member in the house should be expected to reduce expenditure so that money could be saved. Children's expenditure could be checked by simple observation but a homemaker needs budgetary techniques to check household expenditure. Thus, control is according to the performer and work focused.

- **Forward Looking:** The emphasis of control should be on correcting the deviation and helping in achieving goals. Thus, it should be forward looking. For example, when deviation is found in the performance of any member then immediately the focus should be on correcting them rather than blaming him/her or situation for the deviation. Only a forward looking control is helpful and effective.

- **In Built Flexibility:** Control is built in a plan and plans are for future, which is uncertain. Therefore, flexibility in control is essential for meeting unforeseen circumstances, failures or changed plans so that it is still workable.

- **Control Should be Economical:** The cost of control should be worth its benefits. Control should be economical in terms of use of both human and non-human resources. A housewife should spend her time, money and attention to control according to the benefits she can reap out of it.

Thus, controlling is as important as planning. How much the organisation achieves, depends as how much is being applied. Although planning must precede controlling, plans are not self achieving. The plans only guide homemakers in the use of resources to accomplish specified goals. The activities are then checked and controlled to determine whether they conform to planning the action in the use of their resources. Thus, control measures performance against goals and plans and wherever the negative deviations exist, corrective action is being taken in the controlling stage. This will help to ensure the accomplishment of set goals.

Evaluation

This last function of management basically sees whether the goals of the organisations are achieved or not, as planned. It consists of looking back over the steps of planning, organising, leading and controlling to determine as accurately as possible, how good a job has been done. Evaluation looks consistently towards both the process and the accomplishments or results. It is checking and testing of work to tell whether the results are as planned or not. Thus, it measures the effectiveness of plan and judges the quality of results.

Evaluation determines whether the plan needs to be changed and why. It also checks whether effective controls are applied or not. This provides a guideline for the future planning and control measures.

Evaluation is similar to checking in the control step but checking is a quick step by step appraisal of a plan in action, while evaluation, as a separate function in management, involves a complete review of what has already taken place with a view towards better management in future.

Characteristics or Purposes of Evaluation

Evaluation usually tends to help one get away from patterned thinking towards oneself and others, to see a situation freshly. If evaluation from more than one source is summarized and interpreted, we can better see situation in the light in which they appear to others. Lewin sees four purposes in evaluation and its purpose indicates its characteristics:

- To see what has been achieved.
- To serve as a basis for the next plan.
- To serve as a basis for modifying the overall plan, and
- To gain new general insight.

Methods of Evaluation

Similar to control, evaluation can also be done with the help of techniques. The most simple and useful techniques for the homemaker are discussed here.

Budgetary Methods: Similar to controlling budgetary methods, budgets can also be used for evaluation. The type of budgets which can help a homemaker in evaluation are following:

> **Profit and Loss Account:** These accounts help a homemaker in assessing her monetary results at the end of a specific period. If the balance sheet shows the profit or positive deviation it means that she has managed to save whereas loss points at the negative deviation in the accounts. This can help her in evaluating her achievements.

> **Net Worth Statement:** This statement is also used to evaluate the results. The increase in the net worth shows the rise in the standard of living of the family. However, if the net worth of a family is decreasing, it indicates the lowering of standard of living for that family.

Non Budgetary Methods: Like control, evaluation can also be done by methods other than the budgetary ones. Some of such methods useful for the homemaker are following:

> **Summary Reports:** These are the descriptive reports at the end of the execution of the plan to assess qualitatively the results of the process. It shows the quality of goals

achieved i.e. their worth. The homemaker needs to tell others whether the job done by them was good, good enough or not good at all with the help of these reports. She also uses these reports, mostly orally, to point out at the mistakes and to suggest the improvements.

➢ **Checklists:** They are used for evaluation as they are used for control. However, for evaluation, checklists are used at the end of the process to assess the results. For example, a homemaker can evaluate with the help of checklists the extent to which the activities were performed and how much of a goal has been achieved.

➢ **Personal Observations:** Similar to control they are used most often by the homemaker at the end of the process to determine the results or the level of satisfaction it has brought.

➢ **Records:** Household records help homemaker in evaluating the results as they help her in controlling the performance. They can be both short term as well as long term records. These help in evaluating the total economic performance of the family and assessing its net worth.

➢ **Other Techniques Like Audit and PERT:** Though these methods are not used by homemakers but are very important evaluation techniques, usually applied in bigger and more complex organisations.

So we conclude by saying that evaluation is the generally accepted principles that everyone should strive to secure increasingly satisfying results with the resources at hand. It is the core of management—a specific device toward that end.

From the above discussion, it is clear that the homemakers use all the resources of the family, its functions, equipment, information as well as its members to attain their goals. These functions cannot be separated into watertight compartments. Every homemaker uses these functions in varying degrees for the success of management.

At the end, it can be said that all these functions are closely interrelated. But it is essential to treat them as a separate process for the purpose of explaining the detailed concepts important to the manager. At times, these functions are considered jointly in order to show their close interrelationship. For example, organising and controlling can be considered together in system planning. Organising, communication and staffing can be viewed together in studying organisational behaviour.

Thus, it can be concluded by saying, that every homemaker as manager performs the following five basic functions:

- Formulated the objectives and the course of action to achieve the desired objective i.e. plan.
- Establishes the structure, the system and the procedures to operation to achieve the objectives i.e. organise.
- Make the men, machine and methods in proper positions for accomplishing the overall objective and then implement and transmit the plan into actual action i.e. co-ordinate and lead.
- Check and make corrections as and when required as the plan is in action to bring performance in time with the actual plan i.e. control.

- Review and evaluate the complete plan and the control measures applied to determine success or failure through its accomplishments and also to suggest methods of improvement in future planning and controlling measures i.e. evaluation.

Thus, effective management requires the understanding and implementation of the principles and process involved. It is clear that a homemaker can be successful only when she practices them properly along with the co-operation of all the family members concerned.

4

Motivating Factors in Management ≪≪≪

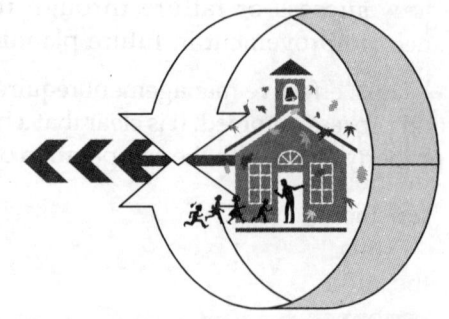

I T IS a well-known fact that motivation is an important factor in all human activities and endeavours. It is not therefore surprising that **motivation** has been defined as an influence of forces that give rise to human behaviour. It has been seen earlier in this book that various factors influence such behavioural pattern and the needs of an individual are the most important among them. Needs not only initiate but also sustain the forces of behaviour and thus seen in almost every human activity.

It is generally observed that every human activity is diverted or moved towards the attainment of the **end points** or the **goals**. They ultimately act as the motive in the fulfillment of these goals. According to Berelson and Steiner the term motive is an inner state that energizes, activates or moves (hence motivation) and that directs or channelises human behaviour towards the'r goals. In other words we can say that **motivation** is a general term that can be applied to the entire class of drives, desires, needs, wishes and other similar forces. While facing all these forces, and in the use of resources towards the attainment of goals, management comes into effect. In all these ways, motivation plays an important role in management.

MOTIVATION IN MANAGEMENT

We have already noted that management is universal. It is practiced everywhere, including in a home. However, we do find a bit of deviation when we look at the factors that lead or motivate our activities in these two areas of management. Whatever may be the deviation, we have to admit that there is one common factor i.e., what we manage are, the people and the other resources made available to the manager. However, when we manage a home or any other organisation, we are not managing just a department or an organisation but primarily the people who compose the department, or the people who are part and parcel of the organisation. Managing requires the creation and maintenance of an environment for the performance of individuals working together in-groups, towards the accomplishment of a common objective.

This environment actually means the creation of such factors, which would motivate the workers to work together. A manager cannot do this job without knowing what motivates people. The necessity of building motivating factors into organisational roles, the staffing of these roles and the entire process of directing and leading people must be built on the knowledge on motivation. Similarly, while managing a home, a homemaker is also concerned with people and hence has to allot duties to its members. In addition, she also has to find out the needs and wants of every members of the family to create an atmosphere for better work performance, and higher satisfaction.

Thus the job of the homemaker is not to attempt or to manipulate people but rather to recognise the motivating factors in designing and creating a congenial environment for their best performance. For example, a small incentive or a gift, an opportunity for further progress or a recognition for good work done by a pat on the shoulder would certainly motivate any one to work harder.

The basic element of human behaviour is some kind of activity, whether physical or mental. We can look at the human behaviour as a series of activities. The question arises as to which activities human beings will undertake at any point of time and why. We know that activities are goal oriented, that is, we are motivated to do things that lead to the accomplishment of something, which we may term as the goals.

The Need-Want-Satisfaction Chain

We can now look at motivation as involving a chain reaction, starting out with felt needs, resulting in wants or goals sought, which give rise to tensions, that in unfulfilling desires, then causing action towards achieving goals and finally satisfying wants. This chain of actions and reactions is illustrated in figure 4.1.

Fig. 4.1: Needs-Wants-Satisfaction Chain

The Hierarchy of Needs Theory

Abraham Maslow[1] gave the most widely referred theory of motivation, which is **hierarchy of need theory**. He saw human needs in the form of a hierarchy, starting from the lowest or the basis to the highest or the most desired.

The basic human needs identified by Maslow in an ascending order of importance are following:

1. **Physiological Needs:** These are the needs for satisfying human life itself like food, water, clothing, shelter, sleep and sexual satisfaction.

2. **Security, or Safety Needs:** These are the needs to be free from physical dangers and fear of loss of a job, property, food, clothing or shelter.

3. **Affiliation or Acceptance Needs:** Since people are social beings they need to belong, to be accepted by others.

4. **Esteem Needs:** This is people's need to be held in esteem both by themselves and by others.

5. **Need for Self-actualization:** It is a desire to become what one is capable of becoming i.e. to maximize one's potential and to accomplish something.

According to this theory of Maslow, these needs follow a hierarchy. Physiological needs are necessary to maintain life and unless they are not met, other needs would not motivate people. After these are met, people are motivated by security and safety need, followed by others till they satisfy their highest need i.e. need for self-actualization.

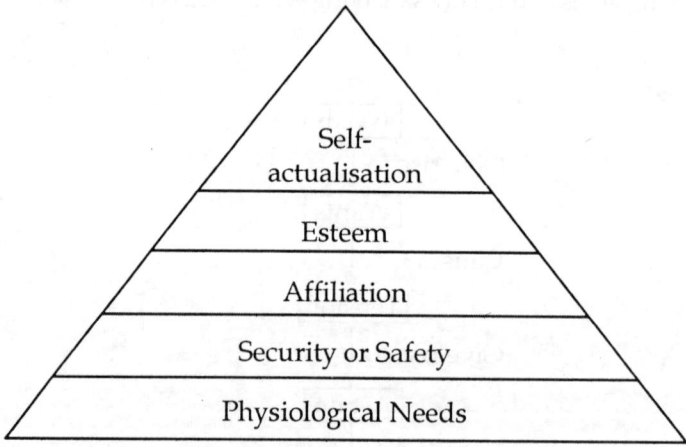

Fig. 4.2: Hierarchy of Needs given by Maslow

The Motivator-Hygiene Approach to Motivation

This theory given by Herzberg is a modification of the above theory. He proposed two-factor explanation of motivation. The first group has needs which are **dissatisfactors** and not motivators i.e. if they exist in environment in high quality and quantity, they yield no dissatisfaction. At the

---

[1]Maslow, A. H. (1960). *Motivation and Personality.* New Delhi; McGraw Hill Co.

same time their existence does not motivate but their lack of existence would result in dissatisfaction. The needs in this category include working conditions, status, security, interpersonal relations etc. This group of factors are called maintenance or hygiene factor.

Herzberg listed in the second group the **satisfiers or motivators**. These include factors of achievement, recognition, challenging work, advancement and growth in job. Their existence will yield feeling of feelings of satisfaction or no satisfaction. Also for these motivators to be effective, it is necessary that hygiene factors are also present. A comparison of Maslow's and Herzberg theories is shown in figure 4.3.

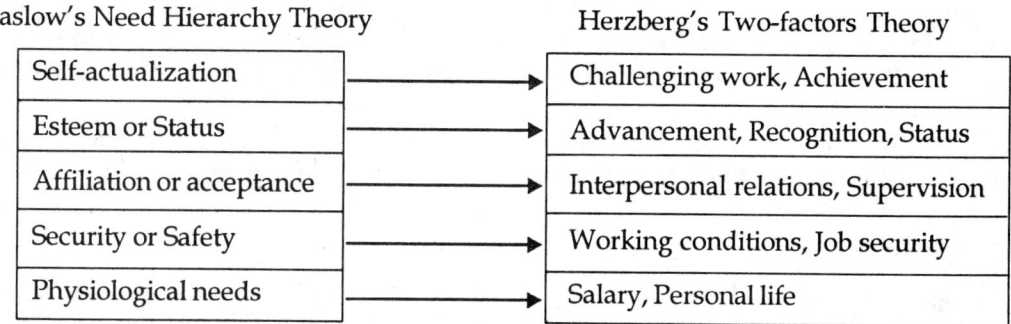

Maslow's Need Hierarchy Theory   Herzberg's Two-factors Theory

Self-actualization	→	Challenging work, Achievement
Esteem or Status	→	Advancement, Recognition, Status
Affiliation or acceptance	→	Interpersonal relations, Supervision
Security or Safety	→	Working conditions, Job security
Physiological needs	→	Salary, Personal life

Fig. 4.3: Comparison of Maslow's and Herzberg's Theories of Motivation

FACTORS FOR MOTIVATION

In reality motivation is more complex than what it appears to be. Individuals and families manage their personal affairs in this constantly changing world. They make decisions to deal with these changing or any other problem situation. The things that underlie all these decisions can be termed as motivating factors. These are

- Values,
- Goals, and
- Standards

Their realisation is the purpose of management i.e. motivation for all our activities and thus the basis for our managerial decisions. An understanding of these motivating factors will help a great deal in all the managerial activities. For the success of management, the manager should be aware of the fundamentals of these motivating factors, i.e. values, goals, and standards.

Values

Values are the key to all motivating factors in human behaviour. Value, as a concept is vague and subjective although it is very important to an individual. Values provide a basis for judgement, discrimination and analysis and it is these qualities that make intelligent choices possible between alternatives. Thus, values are the fundamental forces that force or motivate human activities and endeavours.

Values grow out of human interests and desires. They are the ways behind human actions and the basis for setting goals. They are the products of the interaction between an individual and some object or a situation in his environment. As we seek solutions to problems, values stand as some measure of worth, which is seen as significant in human choices. Our decisions are based on values and our values are identified by the choices we make. Thus values are like a stimulus-response reaction. Some of the values like looking at a sunset, listening to the music, going to see a play or cinema, going to a fair or sitting and chatting with friends give many of us a great deal of pleasure.

These things are the **stimulate** and our **response** is one of enjoyment or a sense of well being. Hence the value found in these experience is pleasure or satisfaction. Thus, we can find that there are a number of values, which are observed in an individual's life.

According to Gross and Crandall *a value is always important to the person who holds it. It is desirable and satisfying. It has the ability to develop in self-creative way and it tends to endure. It is a concept of the desirable, explicit or implicit which governs our choice of methods, modes or goals.*

The term **value** signifies the meaning or definition of worth that is attached to any object, condition, principle, or idea. It is the believed capacity of something or somebody to satisfy a human desire. It is attached with something important or significant. It is something that is considered as valuable.

Parker[2] lists the following as the motivating values of human behaviour-comfort, health, ambition, love, desire for knowledge, technological satisfaction, play, art, religion etc.

Classification of Values

Several authors have reported several systems of classification. In this context, we shall classify values in two kinds with a view to their contribution to understanding and establishing a hierarchy of values in the field of family resource management.

Here, by hierarchy we relate or mean a scheme of relative rank or importance. If conflicts among values develop as a result of their varied origins, such a hierarchy is useful in determining which value will take precedence in a given situation. Without such an order of importance, action might be virtually impossible or the resource may be misused or wasted.

Accordingly we classify values as:

> Intrinsic or instrumental, and
> Factual or normative.

Intrinsic and Instrumental Values: In the first set of classification, an intrinsic value is one that is important and desirable simply for its own sake. It is worthy of being sought for itself alone. Art and its interest in beauty are an example of intrinsic value. Honesty, co-operation, creativity, discipline, respect etc. are some of the intrinsic values in management. Mckee says, the basic intrinsic value of home management is *the healthy growth of individual persons in their multiple relationships in a home situation.*

On the other hand, instrumental values are ways of reaching intrinsic or end values, sometimes called **goal values**. Therefore, they form the basic values leading to another. In home management there are four instrumental values - planning, skills, order and efficiency. Another example is technological satisfaction, which is primarily an instrumental value because it is sought primarily as a means to an end.

Factual and Normative Values: The other classification of values as factual or normative brings out the difference between the factual values that exist, regardless of their level of desirability, and the normative values that have an ethical basis. The factual, also called descriptive, generally are based on people's preferences and desires. The normative are ethical values, which carry the

[2]Parker, D. H. (1941). *Human Values.* New York; Harper and Brothers.

idea of **ought** or **right**. Many feel that the source of normative values is the basis of one's philosophy of life and religion or as viewed by the society as **acceptable**. Some examples of factual values are honesty, religion, loyalty, faithfulness etc.

Values in Social and Cultural Context

The concept of values in the social context is pointed out by Rokeach[3], who defined values as *an abstract ideas, positive or negative, not tied to specific object or situation, representing a person belief about modes of conduct and ideal terminal modes.* Values are usually a global belief about the desirable end status; these beliefs seem to underlie attitudes and behavioural processes. Behaviour is generally viewed as a consequence of our values and attitudes. He further argued that values have to do with modes of conduct and end state of existence. To say that a person has a value is to say that he has an enduring belief that a specific mode of conduct or end state of existence is personally and socially preferable to alternative modes of conduct and end states of existence. Once a value is internalized, it becomes consciously or unconsciously, a standard or criteria for defining action, for developing and maintaining attitudes towards relevant objects and situations for justifying one's own actions and attitudes, for morally defining self in comparison to others.

Analysing the Indian concept of values, Singh[4] has proposed that *a value is that which is a measure of right or wrong, good or bad. It represents what is desirable and what is not desired. The concept of values is normative rather than positive in its conduct. Value is an ethical concept, which determines what the man considers he should do.* Thus, values are normative views, considerations and practices by which an organisation, a group or an individual is influenced in the choice of their action.

Values are socially relevant and culturally embedded. Hofstede found that people of a country share similar values and differ significantly from those in other countries. Hence, he concluded that national culture has a major impact on the values of its people. He found that the values of people from different countries differ significantly on four dimensions. They are

- Individualism versus Collectivism;
- Power Distance;
- Uncertainty Avoidance;
- Masculinity versus Femininity.

Hofstede found that children in collective societies are educated towards **we consciousness**, obligation to family etc, whereas, in individualistic societies emphasis is more on **I consciousness** and obligation to self. People from feminine societies stress on relationship, solidarity and resolution of conflicts by compromise and negotiation. However, in masculine societies, stress is laid on achievement, competition, and resolution of conflicts by fighting them out. In small power distance societies, children are encouraged to have a will of their own and parents are treated as equals. Children in large power distance societies are educated towards obedience and parents are treated as superiors. Similarly, weak uncertainty avoidance societies value ease, indolence, low stress and not showing aggression and other emotions. On the other hand for strong uncertainty avoidance societies, show of aggression and emotions are acceptable.

In any culture some values are regarded as more important than others are, and often organized in a hierarchy or a relative order of priority. Values higher in hierarchy are more important, more enduring and are resistant to change. The values with relatively high positioning

---

[3]Rokeach, M. (1973). *The Nature of Human Values.* New York, The Free Press.
[4]Singh, P. (1979). *Occupational Values and Styles of Indian Managers.* New Delhi; Wiley Eastern.

are termed as **core** values and low positioning values as **periphery** values. Core values are more stable than periphery values but the changing of periphery values allows for continuity and change within a value system. This happens as life advances further.

Management Values in Indian Context

Management is related to a country's cultural environment and is reflected in the ways in which managers perform their managerial duties, which are further dependent on their cultural background and their corresponding values. The basic feature of Indian culture is its joint family system and caste system. Family in India is essentially patriarchal, that is, authority rests with the oldest male member who is considered as the head of family and guides all the decisions. Younger members of the family do not have much say even in the matters that relate to them. All the members are supposed to obey the head of the family. Also the membership to a caste, creed or a group is considered permanent. In this context Gautam studied the Sanskrit roots of the managerial function in India. He concluded that high ideals of leadership are part of the Indian environment and great emphasis is laid on personal purity of action and exemplary behaviour of the leader in public. He also added that the Indian social framework, from the viewpoint of structure and dynamics, has been highly individual-centered rather than centered on the corporate whole. A leader is expected to be nurturing in character towards his subordinates and punitive towards his enemies. The nurturing role of Indian leader is highly eulogized as is his personal integrity and exemplary behaviour. However, the post-industrial urban characteristics o, a leader emphasize rationality and adult ego making all employees look up to him for advice.

Traditional Indian Values: Chakraborty studied Buddhist, Vedantic and Yogic psychology, as well as derivative epic and Pauranic literature to distill the values rooted in the deep - structure of Indian culture and society. He summed them up as follows:

> ➤ **Individual Must be Respected:** This is based on the belief that the divine is enshrined in him or her, whether good or bad, older or younger, rich or poor.

> ➤ **Cooperation and Trust:** The divine inner being of all individuals is unity. Deception or deprivation of others is deceptionss or deprivation of self.

> ➤ **Jealousy is Harmful for Mental Health:** Just as cigarette smoking is harmful for physical health, indicating that any negative values such as jealousy, hatred and anger are like double edged swords which injure the communicator of these values as well as the person on the receiving end. All lead to a stressful mental state and thereby detrimental in terms of group behaviour and performance.

> ➤ **'Chitta-siddhi' or Purification of the Mind:** The noble thoughts of compassion, friendliness, humility, gratitude, etc. lead to refine and accurate perception of human relationship, contributing to sounder relationships and happiness in the environment of home or work.

> ➤ **Top Quality Product or Service:** It is primarily a function of the quality of the mind or consciousness of the doer and only secondarily of quality circles or statistical quality control.

➤ **Work is Worship:** The best way to approach the divine through secular life is to offer each piece of work to the Lord. The guiding principle of Bhagvad Gita is *work without seeking rewards*. This results in complete humility and pure feeling of contentment.

➤ **Containment of Creed:** Whether of tangible like money or intangible like praise as it stresses and robs the individual of wisdom.

➤ **Ethico-moral Soundness:** Every action or karma is a cause for subsequent effect. Ethico- moral soundness gives peace of mind and promotes mental health thereby improving work performance.

➤ **Self-discipline and Self-Restraint:** They conserve energy, strengthen will-power, create trust and confer dignity.

➤ **Customer Satisfaction:** Customer is the divine come upon us in human grab.

➤ **Creativity:** Human creativity is an integral component and extension of cosmic creativity.

➤ **The Inspiration to Give:** Giving is more fulfilling, it adds more meaning to work and life and giving with humility is more dignified than petty needing.

➤ **Renunciation and Detachment:** Renunciation and detachment from selfish results or rewards and egotistic demands in the workplace, from the lower unregenerate ego and its vanity and not from duties and responsibilities.

Contemporary Indian Values

However the contemporary Indian management has evolved through the development of Indian civilization and Western influences originating from industrial revolution. The basic features of Indian management today, are as follows:

➤ Ownership and management are not differentiated, they are overlapping, in the first instance. This is primarily an extension of agrarian and artisan base of Indian society. The farmers own and tilt the land while the artisans own the equipment and manufacture pots and pans, baskets and cloths, furniture and carvings etc. The nature of the traditional economic enterprise is such that differentiation between ownership and management is not called for. This points out to the value, that an organisation is a kind of family where the head has authority and responsibility to establish and accomplish the goals set.

➤ Management in traditional Indian context was centered on the family or religious head. Today the chief of an organisation and the head of the institution play this role. The chief in Indian society has **mana** a benevolent spirit where insight makes him all-powerful. He is considered pure and sacred. He is the custodian of what is good and what is desirable. He must behave as a parent- figure, supporting all the members of the (family) organisation accepting good with the bad, competent with the incompetent, seldom differentiating between social and work roles. In Indian organisations, personalized family-like informal relations are preferred to formal work-related ones. This emphasizes the value of personalized relationships at work.

➤ In its traditional style of decision making, the head is the unquestioned leader and manager who manage with the consent of those who depend on him. This consent is obtained through consultation with them. This is an information gathering system whereby the chief collects points and counter points and incorporates them in his own thinking and decision making. Thus the ideal boss is a benevolent autocrat and yet is valued as a good father.

➤ Hierarchy is the key to management coordination in the Indian context. Horizontal peer- level coordination, so typical of Japanese society is almost missing here. Interventions by persons in the hierarchy are often resorted to in dealing with issues involving horizontal co-ordination. Hierarchy means existential inequality.

➤ The emphasis is on lifelong employment and job security. There is a societal assumption that membership in a family, caste or corporate group has to be permanent. Unless employees have job security, they are not sure, stable or motivated. In contemporary management, there is a lot of debate about the efficacy of traditionally valued security and life long employment.

➤ The belief that there is no substitute to learning on the job prevailed. There is a definite preference for job experience as against job knowledge. Nowhere in Indian history does one find any account of administrators receiving professional training. Being a manager necessitated learning on the job. This underlines the value of personal grooming and experience.

➤ Indian management has a short-term orientation. Past is past, future is uncertain; therefore the present is the most important phenomenon for the manager to deal with. In traditional India, because of seasonal fluctuations, floods, droughts and famines, the farmers and artisans had to focus on the present. Coping with an uncertain demand, markets made the long- range economic planning difficult. The approach to economic activity has, therefore, always been short- term and seasonal.

➤ There is over-concern with job identity. An Indian derives his identity from the social group to which he belongs and the occupation/job he holds. Manager is most comfortable if his tasks are clearly stated and his boundaries of work well defined. There is a preference for job specialisation, not job rotation. This characteristic explains the interrelationship of value of ability utilization and self esteem.

➤ Another feature typical of Indian management is centralization of power and authority, and decentralization of activities, often without the delegation of authority or responsibility. Even in organisations, which have decentralized authority, one finds that people do not exercise their delegated authority and power and often refer back to their superiors. Thus authority is more valued than autonomy.

➤ Indian management values individual excellence. It recognizes individuals who make distinguished contributions in whatever field they work. Collective wisdom is a value in Indian society. Teamwork always commands attention. But there is always confusion in Indian organisations between the reality of teamwork, and the practice of rewarding individual performances.

Thus one can conclude that modern Indian management practices and values differ from those preached in Vedas and Pauranic literature. This difference is due to the fact that management is no more centered on spirituality but is influenced by materialistic, competitive, and other forces in the environment in which it is now practiced.

Value Systems and Value Scales

The collection of the values constitutes a value system. These value systems provide an understanding of attitudes and behaviour of an individual or a group of people. Different scholars and researchers have constituted different value-scales to identify the value system of a group of people e.g. managers, supervisors, homemakers etc. They are also important tools for measuring the values of an individual or a group to arrive at individual, social, cultural or even national value system. Some of these value scales are given below.

England studied the values of managers from different countries. To compare the managers from Japan, India, United States, South Korea and Australia, he used a five-point value scale given in figure 4.4

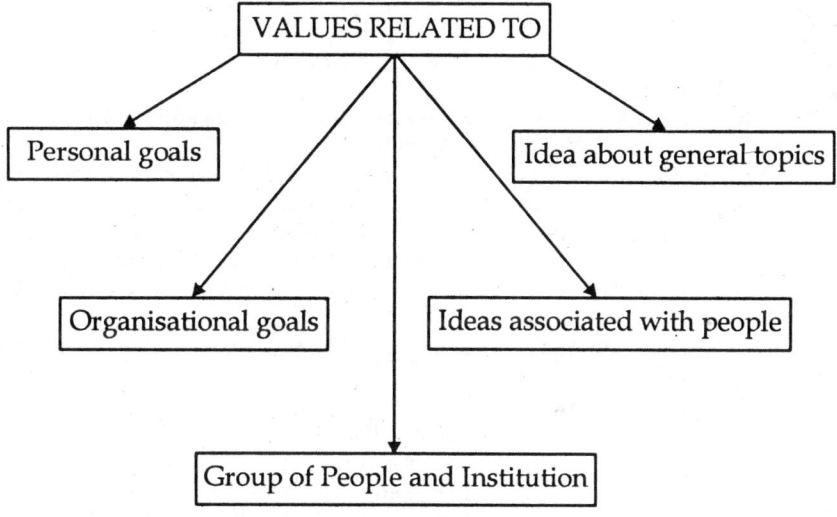

Fig. 4.4: Five-Dimension Value Scale used by England*

Source: England,G.W. (1974). Relationship between Managerial Values and Managerial Success in U.S., India, Japan and Australia.

The value scale, developed to study value orientation by Super is used extensively in many countries including India. Sinha adapted it to suit Indian culture by adding three more values to the original eighteen values. It includes ability utilization, achievement, advancement, aesthetics, altruism, authority, autonomy, creativity, economics, life style, personal development, physical activity, prestige, risk taking, social interactions, variety, working conditions, social relations, peace of mind, physical comfort and dependency. Each value dimension was measured through a set of items. The items were rated on a 4-point scale. Degree of importance ranged from little or not important to very important.

TABLE 4.1: VALUE SCALE

S.No.	Dimensions	Values	Description
A	**Personal development**		
1.		(i) Ability utilization	Efficiency; work output; use of skill and knowledge; foresight.
2.		(ii) Creativity	Imagination; originality; archive application of ideas.
3.		(iii) Physical/ mental relaxation	Sports; outdoor relaxation activity.
4.		(iv) Grooming	Openness to learning; modesty; confidence; good manners; neatness; trust; sharing knowledge with colleagues; leadership qualities; motivation.
5.		(v) Self esteem	Respecting one self, confidence.
6.		(vi) Spirituality	Belief in God; meditation to relieve stress.
B	**Goals**		
7.		(i) Achievement	Task focus; goal orientation; reaching higher standards; job development.
8.		(ii) Advancement	Professional development; promotional prospects; financial and non-financial rewards; technological adaptation.
9.		(iii) Economic reward	Good salary; fringe benefits, perks.
10.		(iv) Life style	Status; prestige; material comforts.
11.		(v) Risks	Adventure; challenge; financial risks.
C	**Interpersonal relationship**		
12.		(i) Authority	Exercise of power; formal and informal control of employees.
13.		(ii) Autonomy	Independence in decision making and carrying out activities.
14.		(iii) Social Interactions	Negotiability; consensus seeking; open communication.
15.		(iv) Social Relations	Interpersonal relations with colleagues, superiors, subordinates; dependency on colleagues.
16.		(v) Self criticism	Reasoning self's failure and taking personal responsibility for it.
17.		(vi) Secure home base	Family stability; sense of belonging; happiness; contentment.
D	**Working environment**		
18.		(i) Variety in work	Job description involving variety in tasks; travelling; meeting different people; undertaking different type of work projects.
19.		(ii) Flexibility	Flexible working hours and ways to carryout work.
20.		(iii) Punctuality	In keeping appointments; finishing work by deadlines.
21.		(iv) Job security	Permanency of job, job notice for quitting on both parties.
22.		(v) Working conditions	Good ventilation, lighting; affluent environment; comfortable surroundings.
23.		(vi) Leisure	Free time for rest or for personal hobbies.
24.		(vii) Aesthetics	Welcoming; attractive; courteous.
E	**Morals/Ethics**		
25.		(i) Social service	Working for the disadvantageous section of the society;. Humanism
26.		(ii) National service	Contribution to national goals.
27.		(iii) Peace of mind	Contentment; job satisfaction: avoidance of disagreement; no conflict or backbiting.
28.		(iv) Honesty/Justice	Truthfulness in work; equity in reward; fair play in work.
29.		(v) Sincerity	Devotion to work.
30.		(vi) Team work	Brotherhood; enhancing harmony amongst team members; 'we' feeling.

Mehra, P. and Seetharaman, P.[5] developed another value scale to study the value orientation of women managers in Delhi. This value scale covered 5 dimensions and 30 values along with their description, as given in table 4.1. This scale was used to study the value orientation by developing a rank order form (ROF) having five groups, one for each dimension. Under each group, two statements for each value were developed. Thus, if a dimension has 6 values on the scale, the corresponding group on ROF has 12 statements. A person whose value system is to be studied, selected the most important 5 statements from each group and ranked them first to fifth as per his preferences. Thus, the data for each group separately was interpreted by calculating their weighted mean scores (WMS), to find the values given most importance by that person under each dimension.

Significance of Values in Day-to-Day Living

Values give meaning to life. In short, they answer the question of why one makes a choice between two or more courses of action. Values are the directive forces. They give direction through the importance placed on objects, situations and ways of doing things with every wish and action. We picturise the thing we wanted and judge it on the satisfaction anticipated before taking any action on it.

Values are the basis for choice. When choices are made at the subconscious level, misinformation and unsound principles may lead to misuse of our resources, resulting in mismanagement. These possibilities call for bringing values to the conscious level, where they will be open for examination and for making the right choices. In the absence of clearly defined values, effective appraisal of practice or decisions is impossible, thereby leading to wastage of time and other similar resources. A well understood pattern or values helps both the individual and the families to define their goals and decide the things of greatest importance to them. For instance an individual valuing education lands up becoming a professional like a doctor or an engineer later in his life, while the person who was not sure, would not have done anything worthwhile.

Values are important to person who holds it. Throughout our life, we are continuously passing judgement either on our own or someone else's, consciously or unconsciously depending on our experiences and the things we desire. We may call them good or bad according to our own individual standards. This is what we attribute as our values.

Changing Values

All values are not stable. Some are continuously changing, bringing a change in the value system over a period of time. Some of the values which have lost their higher importance over time are the following.

- **The Concept of Duty:** Less value is placed on what one owes to others as a matter of moral obligation.

- **Social Conformity:** Less value is placed on symbols of correct behaviour for a person of a particular social class.

- **Sacrifice:** Less value is placed on sacrifice as a moral good, replaced by more pragmatic criteria of when sacrifice is or is not called for.

---

[5]Mehra, P. and Seetharaman, P. (2002). *Management Training and Values: A Study on Women Managing Formal and Informal Organisations Simultaneously.* Unpublished doctoral disseration. University of Delhi, Delhi.

The values, which have gained higher importance over time are following:

- **Expressiveness:** A higher value is placed on forms of choice and individualism that express one's unique inner nature.

- **The Environment:** Greater value is placed on respecting and preserving nature and the natural.

- **Technology:** Greater value is placed on technological solutions to a vast array of problems and challenges.

- **Family:** A high value is placed on family life, but with a vastly expanded concept of family beyond the traditional joint form.

- **Husband-wife Relationships:** A far-reaching shift from role-based obligations to shared responsibilities is visible.

- **Health:** Greater value is placed on one's own responsibility for maintaining and enhancing personal as well as family health.

- **Women's Rights:** A higher value is placed on women achieving self-fulfillment by paths of their own choice rather than through role dictated by society.

The unchanged values are the following:

- **Freedom:** Valuing political liberty, free speech, freedom of movement, freedom of religious worship, and other freedoms from constraints to the pursuit of private happiness.

- **Equality Before the Law:** Placing a high value on having the same rules of justice apply to one and all, rich and poor, black and white.

- **Equality of Opportunity:** The practical expression of freedom and individualism in the marketplace, which helps to resolve the tensions between the values of freedom and equality.

- **Fairness:** Placing a high value on people getting what they deserve as the consequence of their own individual actions and efforts.

- **Achievement:** A belief in efficacy of individual effort: the view that education and hard work pay off.

- **Patriotism:** Loyalty to one's country and dedication to the way of life it represents.

- **Democracy:** A belief that the judgement of the majority should form the basis of governance.

- **Caring Beyond the Self:** Placing a high value on a concern for others such as family or ethnic group; neighborliness; caring for the community.

- **Religion:** A reverence for some transcendental meaning extending beyond the realm of the secular and practical.

- **Luck:** A belief that one's fortunes and circumstances are not permanent and that good fortune can happen to anyone at any time.

- **Success:** In 1970s success for women was seen as getting married, raising children, and owning a house and an automobile and backing the men on their way up the social ladder. In 1980s, get a good education and work hard were two answers, which dominated all others. In 1990s decent and better standard of living, good health, adequate opportunities for one's children, happy marriage and family life and owning one's own home mattered to the people. However, in 2000s the concept of success is self-defined i.e. defined differently by every individual. Thus not only values but also their meaning also change with changing times as seen in the example mentioned above.

Goals

Similar to values, goals also contribute as motivating factors in all managerial as well as in human activities and behaviour. They are intellectual products or intangible things or objects. A goal is more specific than a value and hence can be easily defined and understood. It is an **objective** or **purpose** to be attained and towards the achievement of which the policies and procedures of the programme are fashioned. They are the specific ways of realizing the values, which one holds. For example, if one values efficiency, one of the goals would be to have a well planned house especially the activity areas like kitchen where functional storage and working areas are clearly indicated and provided with utmost precision.

Simply stated, goals are nothing more than the ends that individuals or families are willing to work for. Once the goals are achieved, the individual ceases to work for it. Goals are more definite and concrete than values because they are to be accomplished. As said earlier goals are the tangible things whereas the values are intangible.

Goals also refer to objects, ends and purposes. Certain goals that a family usually goes for are providing nutritive food to the children, the all-round development of individual members of the family within their possibilities and limitations, the development of satisfying relationships within the family or other intimate group. Throughout life, every individual is constantly seeking some objective or goal trying to achieve something in the future. Many of these goals are immediately attainable, while some may take longer course of time. There exists a dynamic nature of goals that one goal stems from another and leads to another one, and so on. Thus it is a continuous chain.

In general, there are three most familiar types of goals that are found while utilising the resources. These goals are

> Short term or intermediate goals
> Long term or ultimate goals
> Means-end goals

Short Term or Intermediate Goals

Families often set intermediate goals or short term goals for themselves. They are the means for achieving long term goals. These goals are more definite than long-term goals and it is easier to form a clear-cut picture of them. They frequently involve decisions or choices made, often unconsciously, from among several alternatives. It is so because they appear and prove to be the best means of achieving some or most of the long-term goals. For example, a short-term goal for a woman may be to clean her home regularly or going to buy things like vegetable and provisions for her family. These short term goals will ultimately help her in achieving long term goals such as to make her home presentable and attractive or to provide variety in the meals for the family. We can thus see that a woman often sets short term goals which also aid in the achievement of long term goals.

Long Term Goals

As the word suggests, the long-term goals are sought over long periods of time and consequently, are ever present. Thus they are considered as fairly permanent. These are the goals that have real meaning to the family group. Although they may be the first goals the family formulates, but they are usually the last the family realises or achieves. In the process of attainment, they set up and realise many short term or intermediate goals because they initiate, determine and influence many of the intermediate goals. They are of great importance. But they are also more complex and require longer time to achieve. Their realisation requires a combination of many activities involving both long-term as well as short-term goals. Educating the children, buying a new car and taking a family vacation each year are a few examples of long term goals. They involve the setting up of intermediate goals such as seeking admissions, saving and allocating money for the purchase of car, family assets or tickets, respectively.

In general it can be said that there are a few long-term goals held by all the families. They are

- Well rounded development of individual members of the family.
- Development of satisfying relations within the family.
- Acceptance of mutual responsibility of the family among the members.
- Accumulating a few assets to have a comfortable living standards.

Means-end-goals

They are lesser or the **in-between** goals. They are the decisions made or the steps taken to attain intermediate as well as long term goals. There are many means-end-goals that are ends in themselves and that are reached with a small number of activities or in a shorter period of time. Whipping up an egg for making a cake, writing a cheque to pay bill or cutting the flowers for an arrangement in the living room, are some of the examples of similar goals, which can be termed as means-end-goals. They involve setting up of an intermediary step in achieving another goal thereby making them as a **means to achieve end goals**.

Role of Goals in Family Life

- **Goal Settings is a Continuous Process:** Throughout life each family is constantly weighing values and keep setting goals, while working for their attainment. When one goal is reached, another is set. Long terms goals are set and divided into short term goals and so on. Therefore, many goals are set and at one time some are attained while families continue to strive for another.

- **Goals as Ends or Targets in View:** All activities of families are directed towards seeking ways of reaching and achieving the established goals. Goals become target as they grow out of our desires, philosophies, attitudes and values. For instance, most of us normally seek happiness and satisfying pattern of personal living.

- **Goals are More Easily Defined and Understood:** They arise out of value, but are more concrete and a clear expression of a person's values. Many goals are set in expectation and by reaching them, they bring a satisfying life. For example, when a person values hygiene, keeping a house clean becomes his goal. Therefore, we find that after values are realised, the goals are set and then the goals are reached and finally the values are satisfied. Similar to values, goals are also related to standards. When a family is strongly motivated by its standards, the achievement itself becomes a goal. When a family sets its standard in serving its meals at the table in a formal way, their every day goals will be to serve the meals at the table with all the dishes crockery and cutlery arranged formally.

- **Attainment of Goals is the Essence of Management:** Management works for the attainment of set goals by allocating optimum resources and resulting in the ultimate satisfaction. When a family prepares a budget, it allocates money for various items of budget such as food, clothing, house, education etc. The family may allot more money for education in the month of July as compared to May or June, when the children have vacations. In these two months, the families may either save money for buying books, uniform and payment of school fees, or it may plan for a holiday in a hill station during peak summer months. However, in a family where there is no school going children, the allocation for education will be missing. If there are grown up children, especially girls, the money is likely to be saved for her marriage. Thus, allotment of resources is dependent upon the expected goals to be attained and the family gets more satisfaction by they allotting the resources according to their needs and goals.

- **Goals are Usually Interrelated and Interdependent:** Goals are carried concurrently as a part of every homemaker's day-to-day activities. They play an important role in everyone's life and motivate thought and action of both as individuals and as family members. They make conscious efforts to satisfy their wants and needs, thereby attaining their goals.

- **Goals are Dynamic:** Goals keep changing, as they are neither static nor rigid. For example, a person saves money with the goal of marriage of his daughter in view, but if she may wish to go for higher education, the saved money would rather be spent on her education. Some goals also become unnecessary or useless as the time passes, therefore, need to be changed. Foe example, when a family saves money to buy a vehicle, it may decide to spend it on buying a house, because it may be more beneficial or the father is provided with an official vehicle. Thus, according to the changing situation, goals are continuously reviewed and revised.

- **Goals are Influenced by Social Environment:** Goals are also influenced by social environment i.e. people living in the same environment have similar goals. A family should therefore, be very careful while choosing goals. It should be realistic i.e. the cost, time etc. required to reach it should be realistically planned, so that families do not find it difficult to attain their goals. For example, the parents in a rural area might opt to marry off their children at a younger age than providing them with higher education. It is common in rural society, that children are married early. They may not require to go for higher education, since they may be expected to carry on their family occupations like farming, cattle rearing etc., where they do not require much education.

Standards

Similar to values and goals, the other motivating factors are the standards. Standards are a set of measures of values, stemming from our value patterns, determining the amount and kind of interest we have in something and the satisfaction we receive from them. They serve as a measure or criterion for the measurement of objects, ways of doing things and ways of living as we make judgements. They are what individuals and families will accept as adequate and worth working, because standards are dynamic. They move the individual or the group towards action. Thus they lead to a pattern of human behaviour and determine their emphasis on the kind of their managerial activities and the amount of efforts to put in.

Standards may be called or defined as the mental pictures of what level is considered essential and necessary to make life tolerable. If achieved, satisfaction results, if not achieved, there is a feeling of discomfort. Even if there is a lack of ability or skill to reach a standard, as long as there is a dissatisfaction or discomfort in not reaching it, it remains a standard. For example, one may be unhappy over her lack of facilities in arranging flowers, such as a flower garden or an attractive bowl, or a container. Once she finds a place for growing flowers and buys a beautiful bowl, she would feel happy to have made a flower arrangement. In this way, she would have found a means of avoiding the feeling of discomfort when she could make an arrangement of flowers, and would stop bothering herself in finding these materials. Thus standards show how much input is placed on items and courses of action.

Classification of Standards

Standards are very important to an individual as well as to a family. For an individual, standards act as self-imposed or socially inflicted demands. These standards may be classified in a number of ways. On the whole they can be classified on the basis of and in reference to their:

- Content
- Degree of Flexibility
- Quality
- The person who holds them
- Measurability (quantitative and qualitative)
- Time (conventional and non-conventional)

Content (Qualitative and Quantitative): Content is that class which is tangible. It is capable of being realized objectively, usually in material form, although it may also be the realisation in the mind. The level of consumption by individuals and families, is a content concept, as it consists of the goods and services that are deemed important for living, the family is willing to strive to attain satisfaction of their wants from their consumption of goods and services. For instance, the level of consumption can be aggregate of the food, fuel and other non-durable goods used up, the services of houses, automobiles, clothing and other durable and semi-durable goods utilised and services of human beings used by individual or group in a given period of time. All these things put together in certain quantity and quality make up the **content** aspect of their living.

Since family resources are limited and must be allocated among many needs, awareness in terms of standard of content is valuable in determining what is more important and what is less important for the family.

Degree of Flexibility (Flexible and Rigid): Fixedness or flexibility shows a degree, varying from rigid, through less rigid, to flexible. Fixedness of standards varies and depends upon place and time. They are consciously adopted to suit a given situation. The difference in the standards of a wife's behaviour towards her husband in India or in the United States and the behaviour of the children in the pioneer days or the modern urban days are the examples of variation of fixedness in their behaviour pattern in place and time.

Rigid standards are often associated with social or religious customs and rites, and are therefore imposed on the family by a social or a religious group to which they belong, while flexible standards are those that allow us to adjust our procedure or content according to the demands of the existing conditions. Acceptance of flexible standards gives greater freedom of choice, and life is likely to be more relaxed, relations less strained and anxieties less apparent. For instance, a woman who is style or fashion conscious in public, maintains a rigid standard in her public life. But if the same woman is satisfied to dress less well even carelessly at home, she maintains a flexible standard in private. However, we generally find that families living in cities follow flexible standards as compared to the families living in villages and rural areas where the standards followed are mostly rigid, though simple.

Quality (High and Low): Quality is more easily understood than other classes of standards, as it is usually what we mean when we speak of the standards of another person, or of ourselves, or of a community. Quality can be defined as mental pictures of what is considered necessary to make life tolerable. Quality refers to the character or essence of something evaluated subjectively and is usually expressed in terms of degree or range.

Standards may be considered high or low, good or bad, correct or incorrect, objective or subjective. We evaluate, by some psychic measure, the quality of conduct or procedure, an object, or someone's thinking. For example, we may say that a homemaker who wears well laundered and ironed clothes all the time, maintains a high standard, whereas the one who wears the same dress two or three times without laundering or ironing maintains a low one. Therefore, we consider or rate the standards followed by these homemakers as high or low in terms of their quality. Besides qualitative standards, there are quantitative or objective standards, which can be expressed in numbers and can be quantitatively measured and scientifically established. These are used for volume and mass measurements. This can be explained in the example of two families — one family serving just one side dish along with the main course of rice or chapati in a meal, and the other family serving two or more side dishes in a meal. The second family thus appears to have a higher standard in terms of quantity.

The Person Who Holds them (Individual and Family): Standards can be classified according to the person who holds them i.e. an Individual or personal standards and family standards. Personal standard is held by an individual i.e. they are important to a particular person, whereas, family standards are reflective of standards of an entire family as a unit. In a family, standards act as demand on the group by all or a part of it, or from outside by some segment of the social group. They are dynamic because they direct the move or make the individual and the group to take action.

However, in some ways, standards act as limits on the individuals and the family group. They compel action towards a thing or an achievement that is expected. They overtly show how much importance is placed on the things and the courses of action. In short, standards act as demands on an individual by himself or by the others.

Standards in general are more specific than values or goals but are bound on one's values. They are definitely specific to materials or objects like standards of food or dress or the other areas of behaviour. They set the limit one will accept in working towards a goal. They are more influenced by external factors. It may be more difficult to make out or see a person's goals and values, but his standards will be better seen or observed and more obvious and specific.

Standards need not necessarily be dictated by an external authority. It is the minor conviction of an individual that the things must be done in a particular way, which is acceptable to the individual and the group that is likely to dictate the procedures. Just like values, standards dictate what is right and just. In this way values are reflected by the standards exhibited by an individual or a family.

Measurability (Objective and Subjective): The standards that are measurable scientifically in terms of quantity, as numeric, are called objective standards. There is no element of bias or ambiguity in stating them. For example, a three course meal will include food served as three different course, having starters, main meal and desserts.

On the other hand, the qualitative standards are subjective in nature. They are the description of levels of an individual. For example, a good meal for an individual could be described as a meal having four dishes with a variety in flavour, texture and colour, artistically placed on a dinner table. While for another individual it may just be a 2-dish meal to satisfy the hunger without any emphasis on the other details.

Time (Conventional and Non-conventional): The traditional standards, which are accepted by generations, are still held high by the people in a country is called conventional standards. They arise from the values held high by the society at large. They are often accepted as mannerism or basic of good behaviour. An example of this kind is the opening of a door of a car for women as a mark of respect. On the other hand non-conventional standards are flexible in nature and·change with time. This can be illustrated by the example of the number of use of two-in-one furniture like sofa-cum-bed in the house. These standards allow an individual to choose their own procedures and working style, as seen in the case of working women who may prefer simple but nutritious meals to elaborate meals.

Standard of Living

In the cluster of concepts in the motivation underlying decisions, values, goals and standards, there is a special all pervasive integrated standard, the **standard of living**. It is a combination of many specific standards. The term standard of living is used to describe behaviour of individual and families to secure what they consider essential for themselves. It consists of a combination of pattern of commodities, services and satisfactions. A person considers this pattern in the same way that he looks upon an individual standard, that is as so essential he will struggle to get or keep it or if unable to do so, will experience discomfort. Every family consumes and sets a standard in the use of various goods, services and commodities as a part of their day-to-day living, such as food, clothing, facilities in their house etc. The use and consumption level of all these goods are put together to reflect and constitute their standard of living.

According to Kyrk standard of living is made up of the essential values to be sought, *it is the attitudes or way or regarding or judging a given mode of living*. It is a subjective view of certain objects or facts.

The standard of living as characterised by Bonde is, usually all pervasive in a given group and is closely associated with the level of consumption (content) because as goods and services become available more and diverse, they are incorporated as part of the way of life of the group members. People regard the things of life and the satisfaction derived from them in relation to what is important, what is valued and what can be afforded. The satisfaction derived is thus a measure and what is good becomes a reality which can be attributed to the level of consumption. The standard is thus represented by the available commodities and services that are chosen and made use of. A typical example for this could be illustrated from the way people celebrate marriage.

Factors Affecting Choice of Standards

Families have different standards. This could be possible even when they share their values. Two families value higher education for their children, yet one family only has the higher standards to send their daughter to a college. The different in their choice of standard could be due to many factors. These factors are

- Cost of available resources
- Fundamental values and goals
- Effect on people
- Social acceptability of a standard

Cost of Available Resources: Standards are governed by the family's affordability. One family, which has enough economic resources to spend on daughter's higher education, will be able to have higher standard for education than the family who could not manage enough money to do so.

Fundamental Values and Goals: Standards are related to both values as well as goals. A family which values economic independence of women and has a family goal of making the daughter economically independent, will have higher standard for education for the daughter than the one who wants to just marry their daughter off at certain age.

Effect on People: Convenience or desire of the family members also determines the standards. A family having a daughter who desires to study hard for higher education or work outside home to become economically independent, will have higher standard for daughter's education than the family having a daughter who simply wishes to get married and become a housewife.

Social Acceptability of a Standard: The standards are also influenced by their acceptability with in a social group. Conventional standards are easily acceptable by large number of families in the society. However, non-conventional standards are accepted by only few. A family living in a society where sending daughters for higher education is largely practiced, may hold the standard of higher education for daughter than the family living in a society where girls are usually not allowed to go out of the house unescorted to the school or to work outside.

Changing Standards

Similar to values, standards also change. They are more visible than values, hence the change in them can be observed more easily. The change in standards can be due to the fast changing economical environment, availability of more material goods, growing globalization, advancing technology, influence of mass media or mere convenience of newer standards. Use of labour saving devices like electric mixers-grinder and washing machine, more use of disposable items like napkins, glasses etc. are now widely acceptable. Thus, with changing times new standards are emerging and replacing the older ones.

INTERRELATEDNESS OF VALUES, GOALS AND STANDARDS

Goals, values and standards are closely related concepts. Value is the base and from values stem the other two concepts, goals and standards, although they in turn exert an influence on values and on each other. As seen earlier, values are important to the individual but vague to express in operational terms. The concept of goal is more specific. It signifies something definite towards which one works. A standard is defined as something used as a basis of comparison. It may vary in precision from quantitative measurement to qualitative measurement. For example, if an individual considers comfort as the most important value, she would strive for living comfortably, travelling comfortably or doing work comfortably. To the extent she holds comfort as a value, she might set the standard by which she will get comfort by making others do everything, or get the required equipment to do work efficiently and speedily, or spend more of other resources like money to assure a comfortable living. Interrelationship between values, goals and standards is indicated in Fig. 4.5

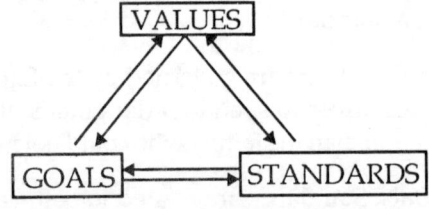

Fig. 4.5: Trilogy of Motivating Factors

In the same way, if a homemaker values health, her goals would be nutritious food and physical exercise which are again dictated by specific standards of nutritional requirements both quantitative and qualitative, as well as the standards of exercise as approved by the health worker or the medical practitioners.

MOTIVATING FACTORS AND MANAGEMENT

Values	- why? (It is a measure of worth)
Goals	- what? (results worked for)

| Standards | - It seems from the values and determine the amount and kind of interest in an item or activity (the way the work is performed) |
| Management | - how? (method of using resources) |

The Fig. 4.6 shows as how values, goals and standards are related to each other and to the management.

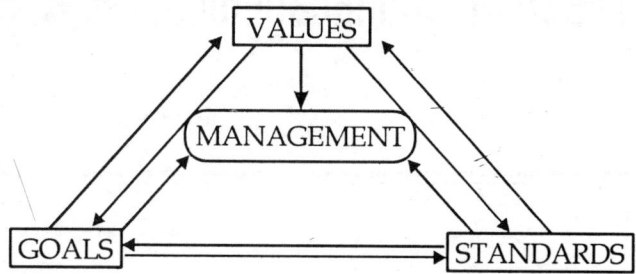

Fig. 4.6: Relationship between Motivating Factors and Management

These three are very important, as family have to constantly make managerial decisions, which are governed by these motivating factors. For example, take value as a motivating factor. It determines human behaviour and helps in decision making and selecting goals and role of management is to attain these goals by proper utilisation of available resources.

Example:

Value	- hygiene of the house.
Goal	- cleaning of the house.
Standard	- how well cleaning is done – use of deodorants and disinfectants

Thus, we use resources accordingly to attain the goal of cleaning the house according to the value hygiene. Thus, motivating factors direct all managerial activities.

In all human activities, motivation is an important factor. It is an influence of forces that develop human behaviour. A homemaker while managing the home has to recognise the motivating factors so as to create a congenial environment for the best performance of all family members. Values, goals and standards are the three motivating factors that form and act as a basis for all managerial decisions. Values provide a basis for judgement, discrimination and analysis. Values can either be intrinsic or instrumental and factual or normative. Goals are the end points set as the objective or purpose to be achieved. These goals can be either a short term, a long term or a means-end goal. Similar to values and goals, the third motivating factor is standards. They move the individual or group towards action. They can be classified on the basis of their content, fixedness and quality. In the cluster of concepts values, goals and standards, there is an integrated standard called as **standard of living**. It is the combination of many specific standards.

These motivating factors, goals, values and standards are related to each other and to the management forming a trilogy. Value is the base and from values stem the other two concepts. Thus, these motivating factors direct all management activities and accordingly the resource allocation takes place.

5

Decision Making—
The Heart of Management

ALL OF us make decisions in our daily life. We make decisions both at home and at the work place. Making decisions is a continuos ongoing process while meeting a problem situation and in solving problems. These decisions bring about a change in a particular situation. In its simplest form decision-making is said to be a process of seeking and selecting alternatives for the purpose of fulfilling the goals. As we have seen in the previous chapters, that whenever there is a problem to be solved or when a choice has to be made, management is involved along with its various steps such as planning, controlling, organizing and evaluating. These steps are basically a series of decisions each based upon the previous one.

The quantity of the decisions made determines the quality of the management process. In this manner, the quality of decisions made determines to a great extent, the effectiveness of a manager. Decision-making is thus the core of all managerial activities. It is in a way, a mental process in solving problems of acquiring and using resources to achieve the desired goals. For effective decision-making certain steps have to be followed in a sequential manner. In a family, a large number of decisions are being made and they may be either individual or group decisions. At times, conflicts arise in a group but that can always be resolved in one way or the other. In general, we can say that decision-making can be said to be the heart or crux of all managerial activities.

WHAT IS DECISION MAKING

Since decisions are constantly being made throughout the management process, decision-making becomes the core of management. It implies that there are some set goals towards which the decisions are oriented. Decision-making is effectively a rational process that involves the selection of a particular course of action from among the various alternatives available, to achieve a pre-determined goal. For this, it is very important that one should be aware of what is involved in decision making, who makes the decision, where, when and how the decisions are made and lastly how the decisions that are made affect our daily life. To have a clear view of what is meant by decisions, a few definitions given by well-known authors are mentioned below. According to J.L. Massic, a decision can be defined as a course of action consciously chosen from the available alternatives for the purpose of a desired result. In the words of R.A. Killion, a decision in its simplest form is a selection among the alternatives. As Gross and Crandall puts it, the making of a decision, or choice-making, is the selection of one action from a number of courses of action, or the choosing of no action. Thus, decision-making is the actual **selection** and no **selection** of a course of an action, when the managerial situations or problems occur. Such decisions made during the managerial process, bring about **change**. However, decision -making is not done all at once, but requires time for its completion. It takes time to predict the consequences of a decision made in a particular situation. It is difficult to suggest solutions to most of the problems. Therefore, it takes time to make a final **selection** or a **decision** to realise its practical applications or theoretical understanding.

Relation of Decision Making to Management

Management occurs when there is some problem to solve, some choice to make. The various steps in the management process are really a series of decisions, based upon our previous experiences. Decision-making is therefore, the heart of the management process. The various steps in the management process are nothing but a set of decisions taken at different stages. Similar to management, decision making is also a mental process. The role of decision making in management involves knowing and actually applying essential information in solving problems of everyday life. Thus, the concept of decision-making is considered by some to be synonymous with management itself because at the moment of choice, the images of the decision maker and manager have much in common.

Simon who is also of the same opinion, treats the two as synonymous. He does not refer to the final choice alone, but to the whole process of decision making itself.

Another approach suggests that decision-making represents one of the steps in management which means it is a part of the process. This would place decision making with planning, organizing, controlling the plan in action, evaluating the results and any other similar steps as part of the management process. It is said that decision making permeates all through the process of management and flows along in each and every step in such a way that decision making may be considered the central activity i.e. the core activity in the management process. The management process, however, is much more complicated than making a decision. Each step in the management process requires many inter-related decisions. A decision is the smallest unit in management like an atom as the smallest particle in a matter. That is why Gross and Crandall calls decision making as the heart or crux of management. Therefore, we can conclude to say that the conceptual framework of the management process involves decision making and decision making plays an important role in the entire process of management as illustrated in figure 5.1.

As the management of resources available for family use proceeds through its sequence of managerial activities or the steps—planning, organizing, controlling and evaluating - we must choose between desires and attainments and the number of choices may be multitudinous. At each step we face problems and situations that call for decisions. Sometimes the solutions or choices that we had thought to be the best alternative at the time it was made, has proved inadequate or deficient. Therefore, before moving on to the next step, new decisions were made. Thus management is diffused with decision making at all times and through time. (Figure 5.2).

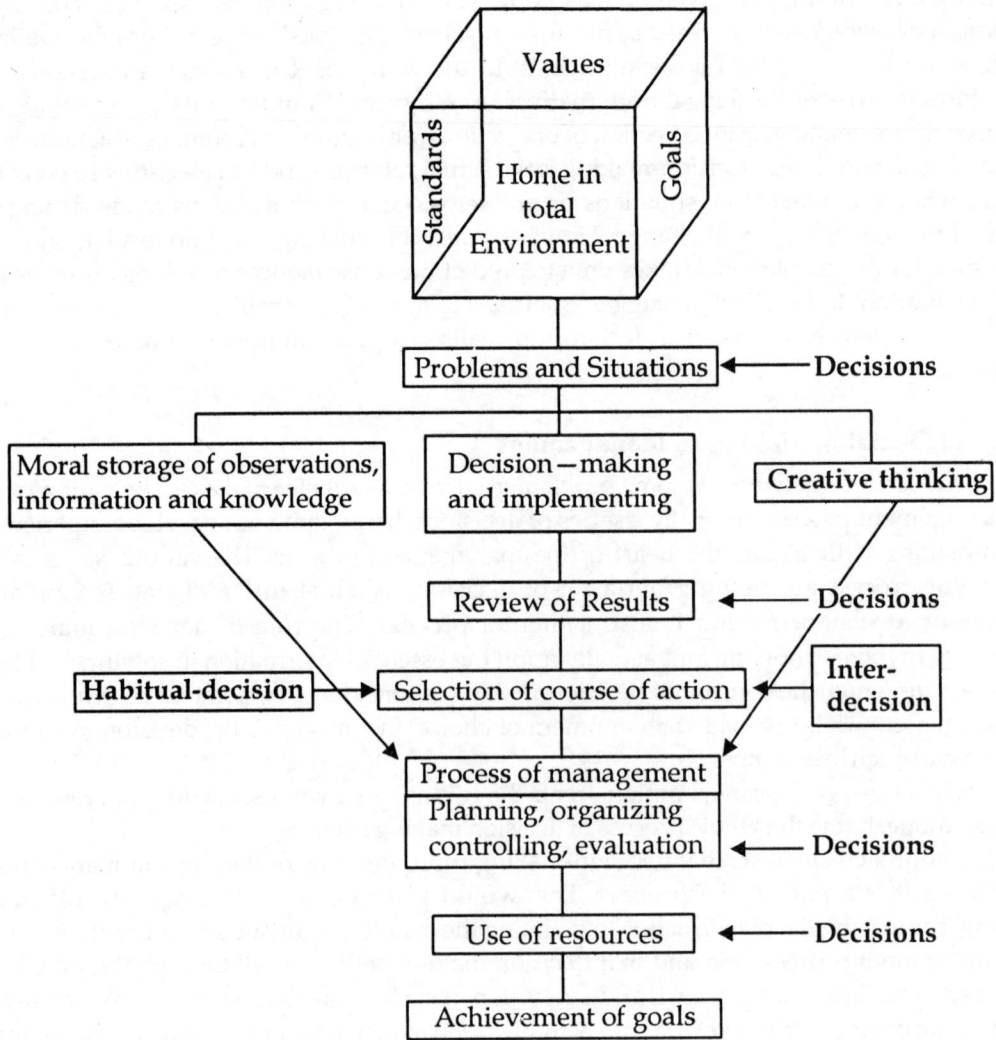

Fig. 5.1: Conceptual Framework on the Role of Decision Making in the Home Management Process.

Figure 5.2 clearly indicates how decision making forms the core or the central part of the management process as one is faced with management problems or situations that need to be changed in the light of the available recourses. One plans an action, or a series of actions, and puts them into operation as he works towards the results anticipated. In carrying out the plan, a decision for each action step must be made as one goes along and reviews progress towards results.

If the problem is simple, the results may be reached in one or two action steps as with *think-as-you-go* decisions. Value judgments are made as each action step leads to the next. When the problem is complex or leads to attending a long term goal, a number of action steps are needed, each demanding decisions. Such steps may take a considerable amount of time. Evaluation of the results of any one action step may show that unwanted sequences outweigh those that are wanted. Under such circumstances a new plan may have to be made or the original one revised, either of which may call for one or more new action steps.

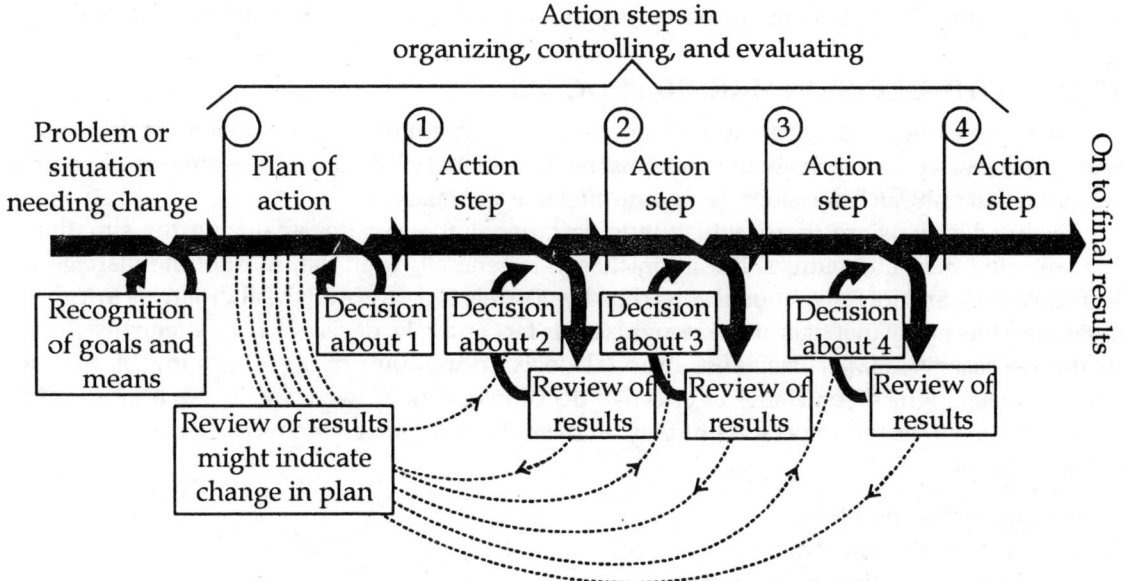

Fig. 5.2: Integration of Decision Making into Management*
*Source: Nickell and Dorsey (1967). *Management in Family Living*.

However, not everything called management action requires decisions. Sometimes action depends upon one's habitual behaviour. Whenever a similar or familiar situation arises, a person simply follows an established *habit*, or a *reflex* even though an *unconscious choice* takes place. All of us follow a set pattern of actions while performing our day-to-day activities. We place the plates, spoons and forks on the dining table without giving much thought to it. We often visit the same shop for purchasing our clothes, groceries etc. We buy the same number of plates, spoons etc. as a dinner set. We can therefore, say that human behaviour results either from *conscious* or *unconscious* processes. While conscious acts involve *true* decisions, habitual acts cannot be regarded as true decisions.

Inspite of the fact that habits do not make up true decisions, they are valuable in management. Simon points out that the use of habits in purposive behaviour permits attention to be focussed on the new elements of a situation requiring genuine decision. The costs of time and energy efforts are high when true decision-making process takes place for every situation or a problem. If habits are eliminated, we cannot imagine how long we will take to do even for a simple task like brushing, combing, eating etc. Even while making conscious choices, like cooking a meal, setting the dining table etc., after a couple of times, many of the so-called decisions are more or less unconscious repetitions of the earlier ones, while working under the same conditions. When the conditions change, as in the cases of the arrival of a baby or shifting from school to college education, conscious financial or other decisions are required, fresh efforts are to be taken in changing the habits. Once again *conscious* decisions replace the *unconscious* decisions in managing family resources.

Thus, decision-making becomes important in home management. It is the way we have to make things happen instead of letting them happen. The making of decisions also implies that there are some set goals towards which the decisions are oriented and carried out. Thus, decision-making is of paramount importance in home management to attain the desired goals.

STEPS IN THE DECISION MAKING PROCESS

We have seen earlier that decisions are made when a problem needs to be solved. At that time a manager draws a conclusion about what must be done in that situation. At the same time, it is also to be seen that only such decisions, which are effective, are taken.

Effective decisions are those, which involve rational choice, after analyzing the situations, problems and their alternatives. But in practice, it is generally seen that most of the managerial decisions made are not fully rational as it is hardly possible to analyze all the alternatives in a given situation. Thus every manager must recognize this fact and take decisions only after considering all the necessary details. Basically, rational decision-making involves seeking of various conscious steps which contribute to effective decisions in their own way. These may to some extent vary from person to person but they all have the common elements as can be seen in the following steps:

➤ What is the problem?
➤ What are the alternatives?
➤ Which alternative is the best?

However, in all problem-solving situations it is necessary to follow a series of steps of phases. Thus the process of decision making has definite and concrete five steps, which are as follows:

(1) Identification of the problem,
(2) Obtaining information about the possible alternatives for formulating the course of action,
(3) Considering the consequence of each alternative i.e. making an evaluation,
(4) Selecting the course of action that seems the best, and
(5) Accepting the consequences of the decision and evolve a feed back.

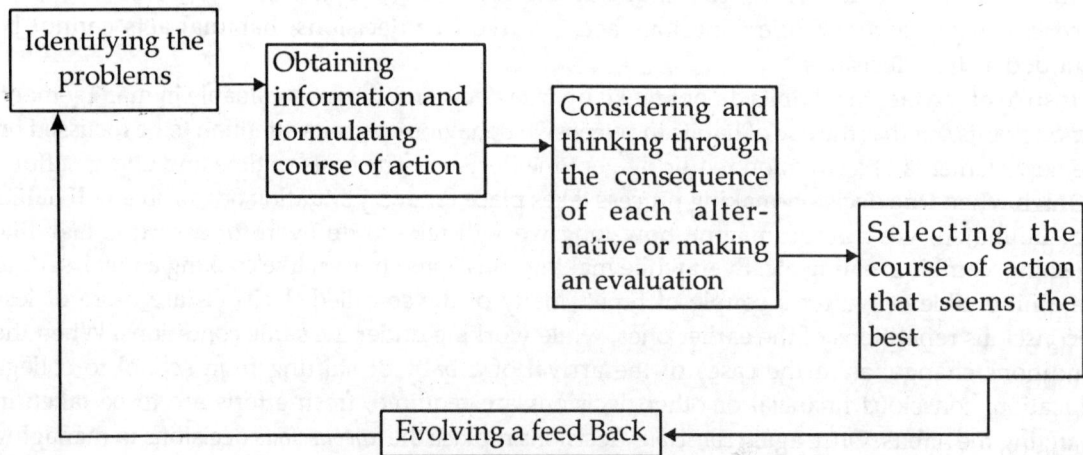

Fig. 5.3: Sequence of Steps in Decision Making

1. Identifying the Problem

Before any attempt is being made to take a decision, it is important to identify the real problem. Thus the problem identification is the first thing to be considered in any decision-making situation. It is the recognition that a problem exists and that finding and defining it is important. One cannot go through effective decision-making process successfully unless there is a clarity regarding the problem. A clearly defined problem is half done, There is no use in getting the right answer to a wrongly realized problem. Many times decisions go wrong because the manager fails to identify the problem itself. The basic reason for such a failure is then the desire to take quick action and to obtain a fast result. As a result, precious time and energy are wasted and a manager may once again have to go back to the original cause because of the hasty decision made earlier.

An example can be taken to underline the significance of this step. Let us see the case of a family which has a trouble in meeting the monthly payments on the house and the housewife thinks that the house is far too costly for a family under these circumstances.

Inability to meet the monthly payments can be due to either wasteful expenditure or the improper use of money such as impulsive buying or purchasing clothing that is costlier than what the family budget has provided. It is also possible that the family did not make any effort to supplement its family income to meet this extra demand. The housewife could have taken up an employment or the adult children could have opted for part-time jobs. So the problem in truth may be either lack of any financial plan or control in the use of the total sum of money available to the family. Therefore the problem identification is an essential factor in any decision-making situation.

For identifying the real problem, the husband and wife must begin to find out the factors underlying the total situation and work to change or correct it before they can act on or settle down. This would not only help them to obtain the desired results, but also help in avoiding the wasteful use of the available resources. To sum up, it can be said, that in the solution of a problem, two things are essential.

> ➤ A desire to find the cause leading to the problem, and
> ➤ The unemotional and unprejudiced view of all the circumstances within the situation.

The family's value system plays an important factor at this stage of decision-making. Individual differences are apparent in problem identification. Age difference also seem to be a factor of concern with problem definition. Therefore, the manager should make efforts to keep in mind these factors, while identifying the problem so that there is no dissatisfaction among the family members as a result of the decisions made.

2. Obtaining Information and Formulating Possible Course of Action

Once the problem or situation that calls for a change has been identified, all available knowledge and information must be used to formulate possible courses of action. This is an essential step for making effective decisions.

The prime objective of any effective decision-making is to find out and select the best possible solution to the problem.

In order to fulfill this objective, he should make an exhaustive list of alternative solutions for solving the problem. The actual evaluation which is the third phase of decision-making process cannot begin unless the decision maker has thought of all the possible alternative solutions to the problem. It can be said that the quality of the final decisions will be proportionate to the number of possible solutions that were considered in the beginning.

To illustrate this step, let us take the example of a housewife who has decided to take up a gainful employment. As a result, she was not being able to do all the household activities which she was doing earlier. Therefore, she has to find out the various alternative sources to get a servant to get the household jobs done. One of the alternatives may be to get a full time servant, or she may employ a part-time servant and request the family members to help her in completing the daily chores. If her family members are also too busy or if her children are too small to do any job, then she may have to buy some labour saving devices. While doing so she has to think of her savings and the expenditure involved, and also to see if she has enough space in her house to keep these labour saving devices. Thus, she has to think of all the possible alternatives before making a final decision, on the choice of an alternative.

3. Considering and Thinking through the Consequences of Each Alternative and Making an Evaluation

After formulating the possible alternatives, the next step is to weigh them according to their merits. This step is essential in the decision making process and is a very crucial step.

The endeavour over here is to determine which is the best solution on the list. This evaluation becomes the process of mentally following through the consequences of each alternative in a systematic way. The evaluation thus involves testing and comparing the solution in the light of certain criteria laid down by a manager. The decision maker, having examined all the information related to the problem, then probes into the future in his attempt to choose one alternative course over another. His objective at this stage is to consider the most promising alternative among all the others, the one that seem most likely to bring about the greatest number of wanted consequences and the least number of unwanted consequences.

Making a list and then comparing these consequences from different courses of action is a way of arriving at a basis of choice. The alternatives should serve as the best means of furthering the families immediate and means end goals with the least injurious effect on the long-term goals.

For the clarity of the concept, the same example cited above can be taken again. As said, in order to finish the household work on time and without getting fatigued, out of the three alternatives suggested, the second alternative, i.e. getting paid help from outside may not be acceptable as the family's resources does not allow them to pay for the outside help. So the family can choose the first and third alternative i.e. working according to the time plan made by the homemaker and also distributing the work to the different family members according to their interest and capabilities. The final outcome of this step would take place in the next step which is often ignored by many people.

4. Selecting the Course of Action that Seems Best

This is the main and final outcome of the decision making process. Under this we usually make a choice by imagining in detail what the results of an alternative would be. This imaginative process leads to deliberation or the weighing of values to discover and choose the best course under the circumstances. Only through such deliberation can we decide which of the several conflicting courses we want to take.

This process calls for the final evaluation because each alternative course of action possesses a unique set of consequences and a final choice is made at this stage. The result of this choice could produce basis for the evaluating criteria for the future decision. Let us once again analyze the example we had discussed earlier. After considering all the possible outcomes, the homemaker decides on the option of taking the help of a part time worker, along with some amount of help from her husband, since her financial position might not allow her either to take up a full time servant or the purchase of some labour saving devices. Thus, this choice becomes the best possible solution for meeting her additional responsibilities.

5. Feed Back

This step, though not always included or realized, is fundamental. Decisions, are made for the future. They are made by human beings to solve their day-to-day problems and thus there is every possibility of their being wrong on certain occasions. Or in certain circumstances, it might happen that we had to modify the decisions made to meet a change in the situation or on account of certain situations arising out of some emergency needs. So, when we are to make similar decisions in the future, our past experience come handy. For this, it is very important to have a sound feed back.

Such feed back should not only help in realizing as to how the decisions are working out in action but also for future needs. In the above-mentioned example, when the homemaker decides on taking help from part time worker as well as from her husband, it might have worked alright for the first few days. Later on she might have realized that it was too tiring and so had to go in for the purchase of a few labour saving devices. Thus, she understands that she should have considered the fatigue factor before deciding on the choice of the available alternatives. Thus feedback would help as a guide in future decisions.

Gross and Crandall consider the last step as *accepting* the responsibility for the consequences of the decision made earlier. Responsibility for a decision rests on the maker of that decision. No *passing of the buck* is legitate. There are usually, if not always, some things about the alternatives not selected that are attractive and desirable. At the same time, something about the alternative selected, are not desirable or attractive. This is likely to happen where there is a lack of agreement between the knowledge and the feelings towards the alternatives selected and towards the alternatives that are not selected. This is generally called **dissonance**. The more important the decision, and the more different the appeals of the various alternatives, the greater is the **dissonance**. When dissonance occurs, there is unrest and therefore, some efforts are made to reduce it. This effort can be explained in simple terms as *rationalization of alternatives* chosen.

FACTORS AFFECTING DECISION MAKING

There are many factors that influence decision making. These factors are:

- Self perpetuating nature of decisions
- Inter-relatedness of decisions
- Time perspectives

Self Perpetuating Nature of Decisions

It is of significance to note this kind of nature of decisions made earlier. If you purchase medicines from a shop, it is likely that you are going to visit the same shop in the next possible similar situation. You will also find similar pattern in the purchase of clothes, groceries, music casettes etc.

The decisions once embarked upon are difficult to change. One may also feel that she can go back and select another alternative, but the *cost* of doing so may be too great to justify the action. Besides, by making the same decision more than once for a particular situation, we tend to develop it into a habit. In that way, we happen to make the decision *unconsciously*, thus allowing the decisions to perpetuate. Thus a person becomes committed to a line of action in such a way that she finds that she must continue it because all other alternatives carry penalties and costs which she would not like to incur. Let us take the example of painting a house. Once the colour scheme for this is fixed, the furnishings and furniture are bought to be in tune with it, thus limiting the choice. At that time the consequences are not felt so much as when they had to buy a furniture item on sale or to use a wall hanging received as a gift.

Inter-relatedness of Decisions

While managing resources, every one makes large as well as small decisions. Larger the decision, the more it will affect our future decisions. A typical example is that of buying a house. Since it involves a huge investment, the allocation for other items of a family budget gets affected. One has to sacrifice a family outing or occasional celebrations with an elaborate menu. Even the purchase of an item of furniture may have to be postponed because of the pending payments towards the housing loan. Thus the choice of buying a home will affect the use of other resources for a long period. Similarly, cumulative effect of many small decisions may affect larger decisions. For example, impulsive purchase of a fashionable dress may affect the larger decision on how much money is to be spent on a jewellery item.

Time Perspectives

The other important factor that affect decision-making is that of time perspective, i.e. looking both backward and forward. Time perspectives vary with different cultures, social classes, individuals and age. The backward time perspective or backward look influences the alternatives found and the consequences expected. Thus an elderly person can sum up her past experiences and feelings in her selections, though there may be many individual differences. There are also differences in the forward time perspective, the emphasis placed upon the future as compared with the present. Experience accompanied by understanding, i.e. knowledge gained plays an important role here. A person choosing a neutral colour, rather than a bright colour coat will find it wearable more often than an occasional use. Thus a decision made at present will affect the way the resources are likely to be used in the future.

KINDS OF DECISIONS IN FAMILIES

While meeting the managerial demands in the home, families are constantly faced with situations requiring decisions and actions. So far we have been discussing decision making as a process. But in a family, majority of the decisions are taken by the parents or by the elders whereas in a way these decisions pertain to all the members of the family.

Some of the decisions related to family activities are:

- Individual decisions,
- Group decisions,
- Habitual decisions

- Deliberate decisions and
- Emergency decisions.

Individual Decisions

As is clear from the name itself, these decisions pertain to one individual only. They are of personal concern. These decisions are more quickly and easily made than the group decisions. One makes these decisions on the basis of one's own goals. They are also influenced by their standards and values. However, the process of decision-making calls for the same number of steps as has been discussed earlier. These include the problem of defining, identifying the alternatives, weighing them and accepting the consequences of the decisions. Feedback also becomes necessary even though it may not be felt. Selection of a particular subject by a student in a school or a college is a typical example of individual decisions, where the application of the entire decision making process is involved.

Group Decisions

Unlike individual decisions, the group as a whole takes these decisions. They are made by the collective efforts of all members of the group each of whom has distinct values and standards. Even though these decisions are difficult to take, they are often made in family situations. They are slowly achieved, the reason being that role conflicts would emerge as each member may consider their expertise superior to the others and sometimes it becomes difficult to reach to a comprise. Thus for effective group decisions, more alternatives need to be suggested and one should try to identify and eliminate the unproductive approaches. At the same time, experiences of many people should be taken as guiding points in the process of selecting the best alternative. When group decisions are made in consultation with the others in the family, there are lesser chances of dissatisfaction resulting in family matters.

Habitual Decisions

These decisions are also called repetitive or routine decisions, which are related to daily activities. These are the lowest level of the decisions, as the action results almost immediately and unconsciously. This is achieved by doing a task continuously or repeatedly. Therefore, it becomes a habitual act. Thus the resultant action is quick and spontaneous. While considering the alternatives too, the choice is automatic. For example, some housewives having the habit of getting up in the morning at six o'clock do so regularly irrespective of the time she goes to bed.

Deliberate Decisions

Unlike habitual decisions, these decisions require longer time and involve the whole family group. In this type, one has to think properly and after careful deliberations only, one arrives at the final decision.

These decisions at times happen in our life. For example, a husband's death in the family may lead to the wife's employment for the financial support of the family. This would be a deliberate decision for the family in order to decide the kind of job she should take. Related to this are the other supporting decisions, which she has to take like her time and energy management and taking care of her family members while she is away to work, etc. Similarly, buying a music system or a television for the family may also require deliberate decisions for the families with a limited income and with small children whose education may be of higher priority.

Emergency Decisions

There are *unprogrammed* or emergency situations that call for decisions that are more complex in nature and require more time and consideration. As an example of such an emergency situation, let us take the case of a family who have lost their home and all their possessions in a disastrous flood. To meet the need for reconstruction or relocation, the family will have to reconsider their use of the available money and plans for investment. This complex situation requires unending decision-making on the part of the family and they may require help from some outside source as well.

Decision Trees

As branches spread out in different directions from the trunk of trees so any complex problems requiring multi-stage sequential decisions are better analysed through decision-tree diagrams. Like the network analysis, decision trees depict the dependence of the main decision point upon all possible future decision-points along with their chance events and probabilities. As a result, decision trees help the managers to see the major alternatives, which are now open to them in the context of future events requiring subsequential decisions. Moreover, the impact of the decision on the realisation of desired result can also be quantitatively assessed from the probabilities of various events incorporated in the tree. For making decisions under uncertainty on the basis of inadequate information, the decision trees provide the management with a means of replacing their intuitive judgment by an objective analysis of the situation. As decision trees show the important elements of a decision and its pertinent premises, they have become a significant technique of decision making.

HOW CONFLICTS CAN BE RESOLVED

We have seen earlier that management in the home involves decision-making. We consider decision-making in management when confronted with problem and then select a specific course of action or a solution from a set of possible courses of action. We have earlier mentioned that group decisions are more frequently taken in a family even though it is a slower process than the individual one.

For making effective decisions it would be better if both the husband and the wife consult each other in all major decisions. This mutual understanding will bring greater satisfaction to the family. Although such group counsels will consume more time and involve the sharing of interests, they are worthwhile when considering their outcome.

An effective decision made more often is a result of the clash and conflict between different opinions and it grows out of the serious consideration of the conflicting alternatives. Role conflicts would generally emerge in the situations, where group decisions are taken. In the home situations, different roles are played by the different family members like the homemaker playing the active role, children playing the passive role and the husband in between these two. Homemaker takes the major decisions, as she is the leader while other members only support her activities. There tends to be more conflicts when these roles are superimposed on the individual. Therefore, it is necessary that some efforts are made to resolve the conflicts. If these efforts are not made, it is likely to disrupt the smooth functioning of the home and result in domestic disharmony and dissatisfaction among family members. Therefore, let us now throw some light on how to meet such situations. There is no single fool-proof solution to all the conflicting situations.

TABLE 5.1: METHODS OF RESOLVING CONFLICTS

	Level of Harmony of Feeling	Method of Resolving Conflict
1.	Highest (complete inner agreement)	Integration/conversion/acceptance of differences.
2.	High	Compromise
3.	Low	Voluntary submission of one side.
4.	Lowest	Struggle and victory of one side resulting in dominance.

Thus these methods of resolving conflicts represent four different levels of harmony of feeling or inner agreement (Table 5.1).

Therefore, when conflicts arise, the different ways should be considered. The family should try to make the best method for these situations. Let us now see how these conflicts can easily be resolved in various ways.

Integration

It is seen that in a family, the conflicts can best be resolved by **integration** in which case whenever there is a problem, satisfying solution is obtained by discussions on mutual consent.

This generally happens in the case of the joint families. An example can be taken of such large joint family, where too much of household work need to be carried out everyday. In order to avoid any confusion regarding the fulfillment of these activities, all the ladies of the house can sit together and then by a general discussion, work can be distributed to all of them by taking into consideration everybody's opinion, so that all these activities flow without any clashes.

Compromise

This is the second best method of resolving the conflict. In this, both the parties go halfway to meet the problem, which results in bringing much satisfaction to both of them. For example, the wife wants to go to her brother's house for a week, but the husband may not agree to this, so both of them compromise to solve the problem. Therefore wife decides to go to her brother's house but only for three days which the husband also happily accepts.

Conversion

This is another method in which the dominant characters explains the situation so convincingly while the other person happily accepts it and is satisfied. A situation again can be taken of highly educated husband and not so highly educated wife, where the husband is the dominant character. The wife wants to purchase a gold necklace to be worn on a wedding, but she cannot purchase that without her husband's permission. Husband does not agree, but he explains very convincingly to her that as he has to repay the loan taken for getting a house, it is not possible for him to fulfill her desire at present. She has to wait till the loan is repaid and when the next installment of money is available to them, they can go in for the necklace. In this way, the conflict is resolved for the time being and at least some amount of satisfaction was achieved which makes the housewife to continue to have some hope so that she is not completely dissatisfied. Therefore, it is better to have a **compromise** rather than not having any hope at all. The resulting conflict is also resolved and the domestic harmony is not disturbed.

Voluntary Submissions

This is a method of *giving in* or of *submissions* to the dominating person like in the above example, the wife drops the idea of purchasing a gold necklace forever but she does not accept it happily. In this case, satisfaction derived is less, though the conflict is resolved. If the husband had opted for a compromise of getting a necklace at a later date, then the housewife would have had at least some satisfaction. Therefore it is better if such decisions are avoided in the family.

Struggle and Victory of one Side Resulting in Dominance

This is the least desirable way for one member to *dominate* over the others, regardless of their feelings. Though one may succeed in this way, but the other person derives less satisfaction. We can take the example of a family staying in a typical rural India where the husband always dominates, he does not want to make any compromise irrespective of the situation concerned. Let us now go into the details by stating the following example.

The wife wants to send her daughter also to school for getting education along with her sons. She wants to do so because she feels that daughters also have equal rights to go to a school and have some amount of education. But her husband does not agree to this and the mother feels unhappy. At the same time, the husband tries to dominate and refuses to listen to his wife's and daughter's plans. Therefore, the husband succeeds and the wife gives in after struggling for some time. Thus, the situation becomes one sided and results in dissatisfaction and frustration on the part of the other person.

COMPLEXITIES IN DECISION MAKING

As we have seen earlier that decision-making is the heart of all managerial activities in every organization, be it a home or a business concern. Decision-making is not done all at once but requires time for its completion. It is not an easy job, even though no practical work is involved in the making of a decision. As quoted by Jacques Barrun that the hardest work of all is to think and decision-making is thinking. In the Indian families although majority of the group decisions are taken wherein all the family members contribute. But still the major responsibilities fall on the homemaker's shoulders. It is particularly difficult for her who has a great number and more kinds of decisions to make with no specialized training in most of them. Therefore, she has to face a number of complex situations while taking the decisions.

It is seen especially in group decisions which are made from the collective action of several individuals each of whom has different values and role perceptions. This is the basic reason for making it a difficult process. It is surprising at times to find a group making a decision without each member being made clear as to what exactly is being decided. This lack of clarity occurs more often when a person is taking the decision individually, and in haste.

It is seen more frequently that effective decisions are not made because the real problem is not identified properly and also because people want quick action. Thus when people proceed too hastily towards a solution, results are not that effective. Thus, to solve this problem, clarity and objectivity are very important. Complexities may also arise as a result of not following all the steps while making decisions.

Another complexity in decision-making is that at times, the manager fails to select the best possible solution to this problem. This generally happens when he does not have a detailed list of alternative solutions. Thus the actual evaluation cannot begin because of lack of all possible alternative solutions being considered to solve this problem.

For effective decision-making, it is desirable not only to select the best possible solution to the problem but also one should explore all possible consequences of an alternative. However, in practical life situation, the person who makes decisions may find it impossible because of either lack of time or of the absence of suitable alternatives to analyze.

Lastly, in case of habitual decisions, which are repetitive type, it becomes difficult for the homemaker to go through all the steps systematically. Thus repetitive decisions are a problem to the homemaker especially when a situation changes. Habit puts an obstacle in evolving a solution more systematically. This is true especially when a similar situation arises, but requires a different alternative.

AIDS IN DECISION MAKING

In Management process, each step involves a series of decisions being made. Decision-making is a continuous ongoing process. Some decisions are just habitual acts but mainly a decision is a choice made amongst various alternatives. Decision-making, thus involves a conscious and rational attempt on the part of the manager to select the best from among the alternatives.

There are various complexities involved in the decision making process, but by following the steps and by working out the solutions systematically, these complexities can be overcome. Besides these, there are other aids that can help a person to masterise the art of decision- making. These are:

> ➤ Understanding the process of change.
> ➤ Changing habits.
> ➤ Providing managerial experiences for making decisions.

Understanding the Process of Change

In general, all decisions involve change. When a decision involves a change, values of a person resist such a change. Normally change occurs in three steps i.e. unfreezing of old elements in a situation, moving to a new level, and then freezing at the new level. People often fail to hold gains in habit changing because they do not realize that freezing at the new level is necessary. Even unfreezing habits is a difficult job. It involves the realisation of a need for change. It also requires the establishment of a situation in which change is expected to take place. It is also important to recognise that there will be some back sliding. Therefore in step 2, i.e. while moving to a new level, one must make greater changes than what she expects to maintain. One of the important aid in long term decisions is understanding the normal changes of the family life cycle. Predictions related to the availability and demands upon the resources can be made more accurate if this pattern is studied in detail. There are many forces in today's world working for and against change. More tension may result when changes are achieved by increasing the forces. However, when the resistance to a change is done at a lower level with less force, same level of action is achieved with lesser tension.

Habit Changing

The past experiences show that motivation is the key point in habit changing. It is not an easy process for anyone to change his or her habits. When the desire becomes stronger, there are more chances for the change to occur with ease. The common techniques for habit changing include talking freely about it, especially to an outsider, and an unpleasant event happening as a result of following an old habit. Another technique involves the conscious repetition of a new habit, and feeling satisfied with its successful achievement. Sometimes, the degree of entrenchment is directly related to the ease or possibility of changing habits. The deeper the entrenchment, more difficult is the possibility of a change. The habits relating to ideas such as ethical norms, cultural values etc. change at a slower rate than the ones related to material things. Habits related to material things change at different rates according to their types. For example, standards for furniture and equipment change more rapidly than the standards for food or clothing.

Considering underlying values is an aid in habit changing. If the change is towards an established value, it is easier to be accepted. For example, if flexibility is a value for a person, it is easier for her to decide upon the menu for her party. However, if the value is for prestige, the same person may find it difficult to decide an appropriate menu for the same kind of party.

Providing Experiences for Decision Making

It is of no doubt that experience is the greatest aid for most of the managerial activities, including that of decision-making. Active participation in the managerial activities by the introduction of a home management house for the students of Home Science in many colleges and universities is a step in this direction. One learns to make decisions by making decisions. This experience should have the following characteristics to be effective in developing managerial ability.

➤ One should make conscious decisions. The actual process of decision- making including the weighing of alternatives must be highlighted.

➤ A person should initially try simple tasks, by realising her own capabilities. Then they should gradually attempt complex ones. Such gradual switch over will aid a person to develop the art of managing situations more effectively.

➤ While handling situations, one should analyse the consequences thoroughly. She should also accept the responsibility for the consequences occurring as a result of the decisions made earlier.

➤ While trying to gain experiences, important issues should be emphasized. Gaining experiences in superficial situations may not actually help a person. Important decisions involving finances, education, career options, etc. are of higher significance than the decisions related to clothes, a day's menu etc.

➤ Knowledge gained in one situation should be used in situations involving the use of similar ones. Analysis of the use of time in one situation can be utilized in the use of time for other purposes. Besides simplifying the task, it can save *costs* involved in selecting the suitable alternatives.

Each step in the managerial process is a series of decisions. Some of them are habitual acts, while the others may be made out of genuine choices among various alternatives, which are true decisions. In a family situation and in other group living, decisions tend to be developed by all concerned. However, opportunities for training and experiences can be powerful aids in making effective decisions. It is for the individual to understand the underlying factors to make effective and fruitful decisions.

6

Resources in Management

RESOURCES are the means to achieve goals. They act as important ingredients in management. Resources are like raw materials, which are processed during the management process. The quality of the result is determined by the decisions made during the process of management, which govern the use of these resources. An effective management ensures the use of family's resources for the greatest satisfaction of the family. The selection and the use of the resources involve the series of decisions and effective decisions help in the maximum utilization of resources. Each time a goal is realised, a family needs to first identify the resources that are available to them, then plan and decide the ways to achieve the desired results through optimum utilization of resources. As resources are needed for every major or minor goal in our daily life, every family must try the optimum allocation of resources in various combinations to achieve all their goals. This is possible when a family has complete knowledge and understanding of all kinds of resources and the amount of resources available to them and the various factors that influence their use to derive maximum satisfaction.

MEANING AND DEFINITIONS OF RESOURCES

Anything that has a capacity to help us reach our targets, can be termed as resource. Resources can be defined as *those material and human attributes that satisfy our wants*. These resources may vary from each other and can be tangible or intangible. They may also vary amongst individuals, communities, states and nations. Meloch and Deacon therefore defined resources as *means, which are available and recognised for their potential in meeting their demands*.

Similarly, Betty B. Swanson also defined resources as *tangible and intangible components which one use to achieve one's goals, objectives and to meet demands*. Gross, Crandal and Knoll on the other hand, also stresses upon their availability by defining them as *they are those available means which are used for reaching goals and meeting demands*.

Deacon and Firebaugh further highlight the importance of resources in a managerial situation by defining them as *they are the supply reservoir for use in the system's specific action and are necessary in some form to solve every management problem.*

Everyone has many resources but sometimes one is not necessarily aware of all the resources at one's command. As a result, we are not able to use some of them at all. Even when we are aware of all these resources, we may not be using them to the fullest extent. However, all types of resources are used to the possible extent to achieve the family goals. They are the 'what' of the management process and enable an individual to achieve his or her goal. Resources are the supply reservoir for use in the family's specific managerial functions and in decision making process.

ROLE OF RESOURCES IN MANAGEMENT

Capacity to Meet Goals

We have already seen that resource have the potential to achieve targets. If we possess money but cannot use it to satisfy our needs, it cannot be termed as a resource. Resource is only that object or capacity that is available to be used for meeting our goals. For example, when a family owns house in another city, it is of no use to them. However, they can make use of the house if there is need for money to meet their day-to-day expenditure. That house will become a resource only when it can generate money to meet their daily needs. Thus, family can convert an idle house by renting or selling it to generate money for meeting their daily family expenses or buy a house where they are living at that time. Therefore, resources help us in achieving goals.

They can be Developed or Generated

Certain resources like human resources, which include knowledge and skills, can be developed. It is also possible to generate material resources like money income, assets etc. with the help of human resource and visa versa. Thus, with efforts we can generate more resources or develop new ones. Development and generation of resources help us in meeting more goals and give us a power to determine the future and feeling of security and satisfaction. For example, a woman has developed a talent of making useful items from the waste material. She can use this talent for decorating her house or can also generate money by selling those items.

Resources can be Conserved or Saved for Future Use

Certain resources like money and material resources can be saved, for their use in future. Thus, they act as reservoir for meeting not only present but also future goals. It is also possible to conserve those resources, which might deplete soon and may not be available in future for use like oil reserves etc.

As resources are so important for us, we also need to identify them. Sometimes we have resources but we are not aware of them. Thus, it is also important that a family is able to identify and recognise its resources. Let us now look into certain characteristics of resources, which are common to all of them and will help a family in identifying and recognising them.

CHARACTERISTICS OF RESOURCES

Some resources are more easily recognised than others. Some are very obvious like money whereas others are less frequently recognised like knowledge and abilities. However, all resources share certain basic characteristics, which are common among them and can also be used to identify less obvious resources. Irrespective of the kind of resource and its source, all have similar characteristics which enable them in reaching goals.

All Resources are Useful

The very definition of resource itself is indicative of this characteristic. All resources have 'utility' which means their satisfying power. Unless we identify the use for an item, it is not recognised as a resource. According to Gross and Crandal, the usefulness or value of a given element may be recognised only in relation to a specific goal. For example, we may cross a road on a number of occasions without noticing a big tree located on the other side of the road. The same tree is recognised and utilised for providing shade when it starts raining heavily or while one is waiting under a scorching sun for another person to arrive. Thus, the location of a tree as a resource is recognised on the specific use or goal i.e. for waiting under the shade and as protection against sun or rain. Thus usefulness of a particular resource would vary for different goals. To cite another example, money is the most valuable resource for procuring admission to a good college, but to do well in a test, one's intellectual ability counts the most. Other resources can support in each management situation to a certain extent. They become the secondary resources in that situation. The usefulness of an object varies according to the problem to be solved. That is the reason why we always identify the primary resource in association with a particular goal in all management situations.

All Resources are Limited

All resources are limited, and some of them are scarce. The only difference in this respect is that some are scarcer than the others. If all resources were in abundance, management would not be necessary at all. The challenge for management lies in the scarcity of resources and still being able to achieve most of the family goals. For this reason we often talk about minimizing inputs of the resources in the management process. The limits, like uses of available resources must be assessed, in relation to specific goals. In general, the limitation on resources can be held both as quantitative and qualitative.

Quantitative Limits

Different resources vary in their limitations and in the accuracy with which these differences can be measured. As said in the chapter on time management, time is the most limited resource quantitatively speaking, since no day can contain more than 24 hours nor can any of these hours be saved for future use. However, the demand placed on time differs amongst individuals. Some perform a job in a short time and some take longer to do the same job. Energy too, is a limited resource, differing in the amount of energy available greatly from person to person. However, energy can be stretched to some extent for future use. If a person does not exhaust himself today, he can do a little more work tomorrow. Latest information points to the fact that the limits of energy may not be as important to some homemakers as are their attitude towards their work. Attitude of a person towards an object or idea might be favourable or unfavourable which gives indication to the limits one can stretch, in making decisions, for completing a task or implementing a change. Work becomes a pleasure when attitude is favourable and people volunteer to assume responsibility. Work undertaken with self motivation and commitment would be more effectively carried out than superimposed work.

The limits of money are easily measured quantitatively as compared to any other resource. Limitations on money is easily and readily recognised too. The amount of this resource varies from individual to individual and the demand for this resource varies from situation to situation and from time to time. Money differs from other resources in that it is limited and at the same time through the investment of human resource, one can procure more of this resource. Money also can be saved and accumulated and if need arises it could be borrowed and given back in the same kind. Other resources if borrowed can be only exchanged for money, since money is the most established medium of exchange.

Similarly, the abilities of family members are also limited resources. The limit is set by the inherent capacities of the individual members but by training, it could be maximized for effective utilization. However, if one lacks intellectual ability, any amount of training may not be effective. The same situation would result when an individual possesses inherent capacity to learn but not able to avail of the training because of inadequate financial resources.

Besides abilities, the material goods a family possesses are limited by two factors, the amount of money available for the purchase and the ability and opportunity for the family members to produce the materials themselves. Therefore, this resource is dependent upon the extent to which other resources such as talent, time, energy and money are available. Yet once possessed, material goods are a resource to us.

Qualitative Limits

While the most obvious limitations of resources are quantitative, there are also differences and limitations in the quality of some of them. This is particularly realised and felt in the area of material goods. For example, one family may possess home furnishings, which are durable, and aesthetically satisfying, while the same family may own furniture, which has neither of these characteristics.

Communities can readily be compared as to the quality of services provided like educational facility or health services. Upto a certain point, the limitation of resources is a challenge, although people vary in the degree to which they accept this challenge. When however, needed resources, which are inadequate, or even completely lacking, there is little reason to struggle to achieve a goal, which is clearly unattainable.

All Resources are Accessible

Resources are those assets, which are accessible for use. An asset becomes a resource only when it is available or within a grasp. Skills of children become family resource only when children are available to help the homemaker. For example if a daughter who is gainfully employed outside the home, has night duty work, is a very skilled cook but her skills cannot be considered as resource as she is not accessible for preparing dinner.

Resources are accessible in varying qualities or quantities. Some are easy to measure and others are difficult. Accessibility of material resources is easy to measure but some other human resources, though accessible, are difficult to measure. For example if a friend goes to the college in a car and gives another friend lift regularly, even friend's car can be considered resource. On the other hand, emotional support and courage are difficult to access, unless demanded though always they are part of human beings.

All Resources are Interrelated

Management in the family deals not only with the usefulness of individual resources but also with their interrelationship. People often have to use a 'resource mix' or combination of resources to achieve family goals. When a young couple wants to decorate their home, they use some of their tangible resources in terms of money, some of the inherited and gifted furniture, as well as some of the intangible resources like innovative ideas and creative ability to paint and decorate the house. Since many material goods are purchased, they are particularly related to the use of money. Goods like vegetables and fruits could not be produced without the input of time, energy and other materials required for cultivation. Time and energy could be conserved if proper labour saving equipment is used to perform a function. Interests and abilities not only determine what goods are needed to achieve the desired goals, but also how effectively other materials would be used. In management, the interrelatedness in the use of resources is important in determining whether or not that goal will be reached. Absence of any resource or too little allocation of one resource may not result in getting the desired product. However, alternatives could be worked out for a single resource in varying limits to be able to substitute for that resource.

Resources are Interchangeable

All resources to a certain extent can be substituted for or interchanged with another resource. In the cases of their scarce availability, or simply for saving money, time, energy or environment, one resource can be used in the place of the other. A material resource can be interchanged with another material resource to save money or energy or even environment. For example if the cost of coffee beans goes up, it can be substituted with tea leaves to save that extra money spent on buying coffee. In general a material resource is easily interchanged with another material resource. A home manager can buy a labour saving device to save energy to save her human resources by interchanging a material resource. Similarly instead of diesel, CNG gas could be used in the public transport vehicles to save environment pollution where a material resource interchanged with another material resource to protect the environment.

A material resource can also be substituted by human resource in case latter is scarce or visa versa. For example if a home manager has little time or energy resource to do household work but has the money resource, she can hire paid help to do the work thus exchanging material (money) resource for human (energy) resource. Reversing the use of these two resources in another situation is also possible. For example, if a home manager has the talent of being a good cook, she can earn material (money) resource in exchange of her human (energy) resource by supplying food for sale. Therefore, it is seen that it is possible to interchange one resource for another resource depending upon their availability, need and situation.

Management Process can be Applied to all Resources

In general, families need to be aware of the potential availability of all resources within the family. Sometimes a homemaker may fail to recognise the human resource available in the family which could be utilized instead of using only the scarce resources like money, her time or energy. Lack of resources may make them feel uncomfortable and poor.

Therefore, a resource gap is felt when we visualize where we want to be and where we actually are. By management we have to narrow this gap by:

➢ getting more resources
➢ making our resources more productive and
➢ changing our standards

For example, let's assume time is a scarce resource. How do we fill this need? This gap between the time we actually have and the amount of time we like to have, can be narrowed by:

> getting more time by starting the work earlier or working longer in the same job.
> Making time more productive by the use of labour saving devices.
> Changing our standards concerning time by using short cuts and convenience foods.

To illustrate this in some other circumstance like washing clothes, it can be accomplished by sending out clothes for washing or by hiring help for laundry purposes, or simply by purchasing a washing machine. Otherwise by proper planning, deciding to wash only undergarments every day, and wash remaining clothes on Sundays or on holidays. Changing standards concerning time can be done based on our judgements. By restricting the kind, the amount and the quality of activities, one can accomplish more.

All resources are manageable to some extent. Their quantity, quality, flow and their use can be regulated to certain degree. Thoughtful planning, organisation, control and evaluation can help a person in selecting the right resource at the time when that resource is in most demand and when it can get the best results. For example, careful choice of savings and investments for the higher education of one's child require a long term planning on how and how much money should be saved every month, besides organising the family efforts, control of money expenditure and evaluation of their activities. All these efforts will ultimately ensure that there are enough funds to cover for higher education. The task of saving money cannot be achieved by a homemaker without the understanding of process of management. She will need to first plan and then delegate various tasks to other family members according to their capabilities and interests so that the required amount of money can be saved. Thus understanding of the process of planning, organisation and delegation of responsibilities along with the authority would be very helpful to the homemaker. Also after this, a control measure is essential to ensure that everything goes according to the plan and if any deviation is found, it is rectified in time. Evaluation at each step and also at the end, can help her in reviewing her efforts and will give her feed-forward for future planning.

Quality of Life is Determined by the Use of Resources

In management of resources, the essential thing to consider is the use of resources and not just the acquisition of goods or resources. In fact, when we consider the level of living of people, we think in terms of the goods acquired as well as the 'use' of the goods. The quality and quantity of imagination employed in utilizing the resources is a determining factor in accomplishing the desired goals. Management and family goals should be considered in relation to each other. Families should gear their management process towards the achievement of their short term as well as their long term goals. Saving for a purpose is more meaningful than saving it for the sake of saving.

It is clear that goals can be achieved only through the use of resources and their management. Therefore, the optimum distribution of resources determines the degree to which a family is actually striving towards a particular goal. Maintaining the health of the family members could be done by proper utilization of money and food resources according to the nutritional requirements of the family members. Time, energy and knowledge are also necessary to fulfill these goals.

Proper utilization of material resources like household equipments can lead to better products in food, clothing and house sanitation which are its contributors to better quality of life. Acquiring knowledge and its proper utilization can help one to make decisions for better quality of life.

CLASSIFICATION OF RESOURCES

Resources are classified in number of ways. The resources that are available to every individual and family can broadly be collected and classified in the different manners. Let us first have a look at some of such classification and try to evolve a comprehensive classification of family resources in the end.

Human and Non-human Resources

The first type of classification is based upon resources within or outside human body. Thus, this classification divides resources into two major categories. They are

Human Resources

   (i) **Personal:** These resources belong to one individual e.g. skills.
   (ii) **Interpersonal:** These develop due to the interaction of two or more individuals e.g. cooperation.

Non-human Resources

   (i) **Economic:** These are productive resources e.g. money.
   (ii) **Non-economic:** These resources are non-productive or for self consumption e.g. cooking for family

Human, Economic and Environmental Resources

The second classification is based on the control of resources i.e. which are within the control of the body of the individual, or outside their body but within the control of the self and the family. The last one are those which are outside their family but can be used by them as their resources. This classification thus includes the following three major categories.

Human Resources

   (i) **Cognitive**: It is the component of thinking using knowledge, setting and defining goals, making plans etc. e.g. intelligence quotient.
   (ii) **Psychomotor**: It assesses the physical costs of work in terms of effects on all the systems of the body that function during work e.g. skills.
   (iii) **Temporal**: It deals with one of the resources with which the family's goals are achieved e.g. time.
   (iv) **Affective**: It concerns the part that personal interests and attitudes play in making the work easy or difficult e.g. values.

Economic Resources

   (i) **Money Income:** They are monetary benefit or gain derived from capital or labour e.g. salary.

   (ii) **Fringe Benefits:** These resources are advantages in goods and services derived as a consequence of employment but exclude money income e.g. paid vacations.

   (iii) **Credit/Elastic Income:** It is the current purchasing power expanded through deferred payments e.g. loans.

   (iv) **Wealth:** It is a composite of holdings, real property and other income producing assets e.g. household durables, equipment, possessions etc.

Environment Resources

   (i) **Physical:** Physical environment includes
      ➤ Natural tangible environment - e.g. soil.
      ➤ Natural non-tangible environment - e.g. air.

   (ii) **Social:** These resources include
      ➤ Social organisations - e.g. NGO's.
      ➤ Economic institutions - e.g. banks.
      ➤ Political institutions - e.g. political parties.
      ➤ Community facilities - e.g. public utilities.

Economic and Non-economic Resources

The third classification of resources is based upon their economic worth. Thus, this classification simply divides all resources into two categories on the basis of their economic ability. They are:

Economic Resources: These are those resources, which are utilised for production e.g. preparing meals for the generation of income (production).

Non-economic Resources: These are those resources, which are for self consumption e.g. preparing meals the family (self-consumption).

Tangible and Non-tangible Resources

The fourth classification is their tangibility or perceptibility by touch. Similar to earlier classification, all resources are divided into the following two categories.

Tangible Resources: These resources can be perceived by touch e.g. money.

Non-tangible Resources: These resources can not be perceived by touch e.g. skills.

Personal, Family, Community, National and World Resources

The fifth classification of resources is based on the social linkages of the organisation to which these resources belong. Thus, resources are broadly divided into the following five categories.

Personal Resources: They belong to one individual e.g. skill.

Family Resources: They belong to the entire family members e.g. house.

Community Resources: They are available to the community from the neighbourhood e.g. banks

National Resources: These are available in the nation and the whole population is provided with these resources e.g. coal.

World Resources: These are shared by the entire population of the world e.g. services of the international organisations like WHO.

Family System and Household, Near and Larger Environment Resources

The sixth classification of resources is based on ecological approach, stressing upon the interrelationship between people and their environment where these resources are found. Thus, resources are divided into the following three categories.

Family System and Household Environment: These are resources available within the household environment, surrounding the family e.g. house, equipment, private transport etc.

Near Environment: It includes markets, educational facilities, recreational groups, medical facilities etc.

Larger Environment: They include forests, rivers, trains, customs, traditions etc.

Human, Physical and Psychic Capital Resources

The seventh classification of resources is based on consumption economics. In this classification all resources are divided into the following three categories.

Human Capital: They include technology, capacity, motivation and time.

Physical Capital: They include the frequency and amount of income as well as purchasing power, elastic income, wealth and community facilities.

Psychic Capital: It is the degree of satisfaction derived from the expenditure of human and physical capital. It regulates the amount and quality of other resources required in the pursuit of satisfaction by all family members.

Renewable and Non-renewable Resources

The eighth classification of resources is based on their supply. In this classification natural shared resources are divided into the following two categories.

Renewable Resources: They include all those resources which can be replaced endlessly i.e. there is an endless supply. E.g. sun, wind, water, geothermal.

Non-renewable Resources: They include all those resources which can be replaced up to a limited period, after which its supply runs out. E.g. fossil fuels, wood etc.

Human, Non-human and Shared Resources

After studying all the above classification, a comprehensive classification is evolved, and suggested in this book, keeping in mind the resources available to an Indian family. This classification has three categories, human, non-human and community shared resources as can be seen from the figure 6.1. These three categories are discussed in detail in the following paragraphs.

Human Resources

Human resources are the less tangible resources, which originate internally and they constitute the personal characteristics and attributes of the individuals. These are the resources, which are available to everyone as a person, in terms of one's capacity, education, occupational status, skills, attitudes, personal traits, and other personal characteristics. As said earlier these resources are less tangible and cannot be easily determined. However, some are more tangible than the others. These are used mostly for productive purposes. In a way, two important human resources, **time** and **energy** are used intensively for achieving work done in the home. They are limited in quality, as well as in quantity. They are more important to individual, since their utilisation will influence the use of the other human resources like **ability, skill, interest** etc. It is the way an individual utilises her time and energy resource that will determine whether the remaining human resources are put into maximum use or not. Incorrect use of time and energy resource will result in their wastage. For this reason, these two resources and their management are discussed in detail in separate chapters.

Abilities and **interests** also constitute part of the resources, which can be put under human resources. If a person can play a musical instrument effectively or be able to do a bit of gardening, one can achieve some of the family goals related to self development or effective utilisation of leisure time through the use of their human resources. Another example of human resource is in terms of the **attitude** of people. Some individuals are always cheerful and optimistic about the work, which can be accomplished in the future, while others live just for today with a pessimistic attitude. Such an attitude affects favourably or unfavourably towards the utilisation of human resource in achieving positive results for the family.

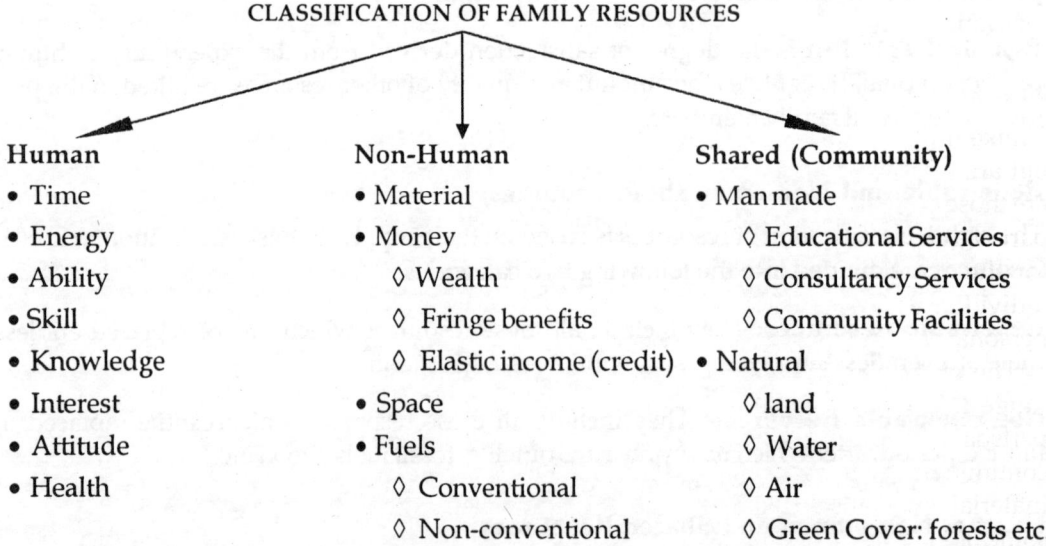

CLASSIFICATION OF FAMILY RESOURCES

Human	Non-Human	Shared (Community)
• Time	• Material	• Man made
• Energy	• Money	◊ Educational Services
• Ability	◊ Wealth	◊ Consultancy Services
• Skill	◊ Fringe benefits	◊ Community Facilities
• Knowledge	◊ Elastic income (credit)	• Natural
• Interest	• Space	◊ land
• Attitude	• Fuels	◊ Water
• Health	◊ Conventional	◊ Air
	◊ Non-conventional	◊ Green Cover: forests etc.

Fig. 6.1: Classification of Family Resources

Another important human resource is the **health** of all the family members. If a family member is sick, other human resources are also not properly utilised and all material and human resources are diverted towards the care of the sick. Thus, the health of the family members, especially that of the homemaker is a very important resource, which also governs the extent that the other human resources can be utilised.

Apart from above mentioned resources, the **qualities of managers** discussed in the chapter 2 can also be considered as human resource of the homemaker. The self confidence, self esteem, information and awareness, communication abilities, leadership quality, team building spirit and synergy of a homemaker also help her in performing her work better. These qualities become her resources, when she confidently leads the others in her family, combines and coordinates their **efforts** (synergy), builds a team with necessary skills, uses and communicates the necessary information to achieve the desired goals of the family.

Although everyone has some human resource or other, one may not be aware of all of them and also when it could effectively be put into use. A woman who is artistic and knows stitching can wonderfully clothe the family in relation to the actual requirements of the family. Thus, she is also able to save money, which otherwise be required for the payment to the tailor. The same is true of people who have a flair of designing interiors. With their innovative ideas, they would do a marvelous job of designing their interiors with the least money inputs. A homemaker who is interested in gardening and has the necessary space, can grow vegetables, which would provide a part of the meals on the limited budget. A homemaker consciously planning meals for the family will make use of human resources, **knowledge** in planning nutritive meals for the family to meet the quantitative and qualitative standards of their nutritional requirements. She would also make use of inexpensive sources of the various food groups if she has to plan all these within a specified budget. Similarly, a person who has the **artistic ability** to make flower arrangement can beautify her home effectively by making use of limited material resources. Also, the person who has the positive attitude towards savings would spend less at present and would keep aside certain amount of money for future emergencies. Some homemakers are so resourceful that they are able to manage very effectively with very little tangible resources. Thus, it can be concluded that realisation of human resources is necessary for the attainment of family goals.

Non-human Resources

Unlike human resources, non-human resources are those, which are external to the individual, but are utilised, possessed and controlled by the individual. Non-human or material resources are those tangible resources, which are available for use by any one and everyone, but they are not part of a self. These are more easily identifiable than the human resources. They are also essential for the achievement of most family goals. However, they are limited in availability to an individual as well as to a family. Non-human resources in general include material **goods, money,** personal or family **possessions** or **assets, land,** working **space** in housing and **fuels.**

The management of **material goods** presents a most complicated problem today. First, they range from food mixers to safety pins, from clothing to pens and pencils and from highly perishable foods to books. In general, each kind of article requires different treatment if it is to continue to fulfill its function. This is true even when two articles are constructed of the some material, for example, woolen blankets and woolen rugs, or when two articles are identified for some use but made of different materials, as in sweaters made of wool compared to those made of orlon.

Moreover, materials are changing constantly bringing new characteristics which may or may not be advantageous and which require different treatments from those to which the family is accustomed. Thus, one has to study the different materials carefully to use them effectively in management situations. Along with these material goods, **money** is also a very important non-human resource. In fact, it is the most difficult resource to manage as it is the most sought after resource. It is limited in quantity as well as in quality. It is a very important financial tool, which helps the family to achieve most of their goals and satisfy their desires. It is an indicator of economic and social status for families. Families should therefore, be very careful while handling their resource. Another resource, which requires functional planning, maintenance and control is space.

A family uses **land** and **space** for residential purposes as well as for economic generation, by building house, shop or factory. Space is an important resource for a family, since it will determine how well or how poorly the family is able to perform their day-to-day activities. Adequate space coupled with good lighting, ventilation and storage facilities will make it functionally effective while performing the household activities. Space is therefore a very important resource, which needs to be tackled and organised properly. Keeping this in view, use of space and its organisation is dealt separately in the chapter on space management.

Similarly, **fuel** is another very important resource used extensively for cooking (LPG, coal, kerosene etc.), for running vehicles (petrol, diesel, CNG etc.), or for lighting or running various equipments (electricity). How a family utilises these resources determine the standard of living of a family. However, families can also use biogas, and solar energy for meeting their fuel needs for cooking purposes.

Shared (Community) Resources

These resources can be broadly categorised into natural shared resources and man made shared resources. **The natural shared resources**, which are at the disposal of all the families include air, wind, water, sun, fuel, fodder for animals, land and green cover. The level of purity of air, availability of clean drinking water, management of solid waste and protection of green cover are fundamental facilities which govern the lifestyle and health of the family.

One city may provide ample housing and, another may not. Recreational facilities may range from the legitimate theatre in some communities to nothing but small inn in others. Less polluted city with ample green cover like parks and forests around it, provides the important community resources, which governs the quality of life and health enjoyed by its population.

Land, which is a very important natural shared resource is used for building homes, agriculture, construction of roads and railways, building drainage, disposal of garbage etc. Thus, roads, railway and transport facilities, and housing facilities are important man made shared facilities constructed on land. These constructions along with cutting of tree for various purposes, mining for metals and agricultural affect the land to a large extent. Thus, individual, family, community and government should take adequate measures to protect land against deforestation, soil erosion and soil pollution. The communities which have good drainage, adequate and efficient solid waste disposal system, green cover etc. provide good natural shared resources to the families living in it. **Green cover** is a 'green belt', or the land covered with trees giving it the shape of forest. Growing more forests or growing trees where forest have been cut is called aforestation and the land under this green cover is prevented against animal grazing and cutting of trees. Thus, it provides relief from pollution and preserve environmental resources.

Water is another shared natural resource used extensively for domestic as well as industrial purposes. Availability of portable water for drinking and other uses is an important shared natural resource which governs the life style and health of the families. The natural water gets polluted by pesticides used for agricultural purposed, disposal of industrial and human waste into the water bodies. Thus, the efforts made by individuals at family level, municipality at government level, industry and agriculturists at their own levels can improve the quality and quantity of water available for use. Thus, the families living in the communities having clean and pollution free water for domestic use, safe and hygienic waste water disposal system enjoys the better quality of this natural shared resource.

Similar to land and water, **air** is also a very important shared natural resource. We all breaths air but the quality of air available for this purpose differs from one area to another. The air having 78% nitrogen, 21% oxygen, 0.03% carbon-di-oxide and 0.97% other gases is termed as pure and pollution free. The toxic gaseous emissions in the air, pollutes it by changing its composition and breathing such air causes from minor to major health problem. Apart from these, the air pollutants like carbon-di-oxide, methane, nitrous oxide and chloroflouro carbons, adsorbs infra radiations causing atmospheric temperature to rise. This has lead to global warming which in turn affects the natural resources of earth by melting ice peaks, raising water levels of oceans and thus, reducing availability of land resource.

Apart from these, **sun, wind** and **hydel** (water) are other natural energy resources which are also renewable. These natural resources can be tapped by a family and community for various purposes as cooking, lighting of the house, heating water, generating electricity etc. These resources are free for everyone in the community, renewable in nature and free from pollution.

The **man made shared resources** include some human as well as material resources such as **educational facilities, health services, consultancy services** like counseling agencies. The resources such as community centres, creche, clubs, libraries, shopping and recreational facilities etc. along with services like roads, fire services and police services are available to everyone living in that community and therefore, are important shared community resources. Thus, these shared resources by nature are of both human and material, they together contribute as community resources. They are made available and provided equally to every family in the community and in the neighbourhood. It is for the family to realise and use them effectively and prudently, as they are to be shared by everyone in the community. The community members not only have a right over them but should also behave like responsible citizens.

Thus it is important to recognise and use all kinds of community resources available to the family for the achievement of family goals. However, all communities differ greatly in the resources they provide to their members.

The first step in management is to identify the various resources and plan their use. One can readily identify those resources, which originate internally and those originate externally. It is equally important to identify the shared community resources which may not get realised or recognised easily. The success of the management process in the home depends mostly upon the recognition, allocation and use of all kinds of resources. In order to do this, one should have the basic knowledge of the nature and characteristics of resources. Identifying resources enables everyone to ascertain the specific resources and their its potential use for accomplishing success in their management.

FACTORS AFFECTING THE USE OF RESOURCES

Though it is possible to apply management process to every resource, there are certain factors which influence their use greatly. These factors which we are going to discuss now are closely related to each other. However, these factors do not operate in a vacuum, and it is very difficult to isolate completely the effects of a single one. An attempt has been made to highlight the effects of these factors.

Size of Income

Money is a very versatile resource. It has an instrumental value and brings the highest personal satisfaction. This resource has a number of alternate uses. It can be exchanged for non-human resources such as material goods and also for human resources such as skill, time, energy, etc. The size of income is correlated with the purchasing practices with decisions such as to when clothing and furnishing should be discarded, with the amount of paid help available to the homemaker, and with the presence of facilities and equipment in the home. When the size of income is more, there is a possibility that the family will be able to attain more number of these goals. Larger the money income, higher will be the satisfaction of the family members. While the size of income affects the use of all resources, most of the information available on this resource stresses the effect of income size upon its distribution on the family needs. A knowledge of this impact help a family to acquire and utilize the other kinds of resources in larger measures instead.

Socio-Economic Status

Social status indicates a perceived relationship of a person to the social group. Social status accounts for differences in family values, attitudes, decision-making and in expenditures pattern. In the modern urban areas, status is something which is achieved and not ascribed through birth. There is a social stratification in the society, as all people do not enjoy the same life style. An individual has many different positions in a society, each of which may have distinct status implications. Therefore, the individual's status is a composite of these different and sometimes contrasting aspects, rather than the result of any one. These aspects and their interrelationships are constantly changing. The complexity of status determination demands great care in the selection if indices to status. Status groups are classified as upper-upper class, low-upper class, upper-middle class, lower-middle class, upper-lower class, and lower-lower class. Ordinarily an individual's use of resources will be greatly influenced by his own status group. Many families consciously or unconsciously, choose upward mobility as their goal. Family is a social unit through which an individual finds his place in the class structure. Class, status and power are closely related terms, which illuminate the many dimensions of social structure.

In division of responsibilities and roles of family members, most authors agree that in lower class families more managerial activities are carried on by the mother than by a middle class mother. Today middle class covers a wider range of people who are fast moving in the social, structure. Upward mobility occurs only in terms of their education and occupation.

To maintain a higher status, a family has to live up to it and this directly governs their expenditure pattern, values, goals and standards of status.

Occupation

Traditionally the family's life-style, which include time schedule, the entertainment schedule etc. has been influenced by the husband's or the head of the family's occupation. The time schedule of the doctors and businessman's family, differs from that of an ordinary middle income salaried family. The other factors such as the size of income, the status of the family, and their social circle, are also to a large extent determined by the occupation of the father. Travelling allowance, housing facilities etc. are all determined by the occupation, which again affects the status and the living styles of the family.

Gainful Employment of the Homemaker

Women in the paid labour force affect resource allocation of all family members. Adding to a family income is considered one of the major reasons for married women for taking up gainful employment. The expenditure pattern also change with the employment of women. Gainfully employed women use more purchased services, spend more on transport, clothes, labour saving devices, fast food etc.

The time pattern of employed women indicate an expected longer working day. They spend less time on household activities and their time schedule also differs from the non-gainfully employed women. She does her work in the morning and evening, thus her energy usage differs. Moreover, adjustments within the family need to be made if the homemaker uses a large part of a time and energy in gainful employment, thereby inc. asing the amount of money available but decreasing her time and energy available for homemaking tasks. Increased income leads to an increase in personal satisfaction, with less psychological fatigue. However, their demand for other family members' help increases. When women work outside the home, children happen to do more of the household tasks. Besides this, synchronization of occupation and household work is also very important, which the homemaker has to keep in her mind.

Size and Composition of the Family

The family size, age and sex make up the family composition, that affect their resource usage. If a large family attends to maintain the same level of consumption as that of a small family, it is clear that it will require more commodities and services. Food, clothing, personal care, medical care and entertainment expenditures vary quiet directly with the size of the family while expenditures for equipment, housing and home furnishings are less variable as the size of the family changes.

The availability of the resources like time and energy for use increases as the size of the family increases. Energy decreases with age, as well as with decreased size of the family. Although the hours spent in various activities increase in large families, the proportion of each activity to the total home making time varies only slightly. If one or more of the family members are an infant or a young child, an extra demand is placed upon the time of the other family members. Similar to the size of the family, in the use of time, energy and other resources the composition of family assumes importance. If a large family is made up of adolescence and adults, more human resources are available and extra demand on money resource may be expected. In a family where there is a grown up daughter, she may contribute to the availability of human resources by sharing the activities of her mother whereas a grown up boy may not be able to do so. Thus the composition of the family is terms of sex affects the availability and use of their resources.

Motivation/Attitude

Motivation is an internal attitude. It is the way people use what they have that is important in meeting and establishing their goal. It directs or limits the quantity, quality, and the mixture of resources that a person is willing to use in goal attainment. When two children of the same family have two different attitudes, the one with increased motivation and positive attitudes than the other child, will do a better job as compared to the other. In a family where there is no motivation of family members with unfavourable attitudes towards the availability and use or resources, it is likely to affect the kind and number of goal achieved by them.

Education

Formal education has direct influence on the amount of income earned. Person with increased education have more likelihood of earning better than others. Women with more of formal education tend to take up gainful employment and so have less number of children thereby affecting the size of the family. Educated people have more geographic mobility, increased income and economic stability. Education is related to sharpened ability to think abstractly - an ability important for planning and decision making, creativity, resourcefulness, etc. Increased exposure to social environment also comes with higher education. As a result, education affects the availability of resources and the way these resources are utilised. On the whole, education determines the quality of life a family is likely to enjoy.

Family Heritage and Cultural Background

This is an influential factor in transmitting family values. In our Indian society, we have joint family system in which grandparents, parents and children live together. The traditions, values, beliefs of the family are transmitted from older generation to the younger generation orally.

In India women's education and occupation are very important. Mahatama Gandhi was right in saying *When the women of the house is educated, then the whole house is educated.* The cultural background of the family also affect the use of resources as the eating and spending, hobbies, their beliefs, festivals, superstitions etc. all are governed by this factor. This directly influences the way family utilises its resources on various occasions or similar situations.

Location of the Family

The location of a family within any community in relation to shopping areas, schools, place of husband's work and so forth will affect the homemaker's use of time, energy and other resources. Families living near the city or near the city, all the community facilities such as markets, schools, parks, clubs, banks, post office, etc. will be easily available to the family that may affect the use of resources such as time, energy and specially the money.

There is some marked difference in the use of their resources between the families in village area from the city families. Village families make up for an exchange of values because of the close, almost enforced association of families. There is an impulse towards culture and good taste in this exchange of values. In a city, families are so busy within themselves that they have no time for socialisation or to think of culture. Whatever they need, they buy them from the market. Also their nearby environment affects the use of resources such as people living in college campus will spend more on education. Similarly, families living a good shopping complex will have the tendency of spending more on clothes, eating out, shopping, etc. and so on. Nearness to or being away from the place of work or a school may affect the expenditure on transport or similar needs.

Health

Health of all family members is important in home management. It is a state of complete physical, mental and social well being. Illness can make increased demand on managerial activities and resources such as money, time and energy. All such major resources are diverted when a family member's health is affected. A family, which enjoys good health, has an increased availability of resources to utilise or to meet more of their needs.

MAXIMIZING THE USE OF RESOURCES

It is now clear that a large number of factors affect the availability and the use of resources. This may reduce the level of satisfaction a family may derive specially when there is limited availability of resources. In order to increase its available resources and to get a higher level of satisfaction, a family must be aware of a number factors. These are presented in the following chapters.

Careful Selection of Goals

It is unfortunate if goals, which might have been easily achieved by using all available resources are laboriously achieved through the use of only a few. For example a woman who because of limited energy, cuts down the expenditure on all other items to provide elaborate labour saving equipment in meal preparation, deprives her family members of some desired goods. Instead, the father could have improvised some labour saving devices and other family members might have offered to shoulder some of the responsibility for meal preparation and the mother might have some opportunities for saving energy by simplifying means and becoming motion minded in preparing them. The entire family could have sat together, analysed the problem, chalked out the ways of helping the mother instead of wasting money on a new equipment. While doing so, they could have also decided together some other goals for which the money spent on the equipment could have been spent.

Positive Attitude

A change in attitude can easily represent an increase in resources available for reaching a desired goal. It is always less easy to increase the supply of energy to pursue other resources. Many a times the young family members are reluctant to do household work. If these members develop a positive attitude towards work, they can utilize their human resources more constructively thereby maximizing satisfaction from the use of these resources.

Interchanging Available Resources with the Scarce Ones

Maximizing the use of resources often allows the substitution of one resource for another which is scarce. For example, if a homemaker has more time than money at her disposal, she does all her household work herself. On the other hand, if she has more money at her disposal than time, as when she works outside the home, she may hire someone to do certain tasks or use labour saving facilities. Also through knowledge, skill, use of convenience foods and an analytical approach to her work, a homemaker can reduce the amount of time required for house care. The homemaker can also save her time by sending her child to a creche. In this way, she is substituting her human resource with a community resource.

Another example of substituting one resource for another is found in a family which recognizes the validity of community's demands upon their resources. They may see possibilities for using their energy and abilities to satisfy family and individual needs in order to free funds to but an additional asset to the family, like the example of painting the kitchen themselves and use that saved money to meet that family's other important needs. Similarly, a homemaker can agree to take a child from the neighbourhood house so that she can utilize her leisure time to earn an income.

Not only a general resource such as money have alternative use, but some specific good also have many possible uses. The selection of such good increases utility such as, a beautiful bowl of oven China can be used for a casserole, for salad, for fruits or for decorating flowers. Similarly a card table can provide a surface for study, table for light meals, a service table as well as its designated use for cards. Satisfaction also increases if one finds new uses or combinations of uses for things already owned. Such as finding that a blouse purchased to wear with pants looks well with a skirt already owned increases the utility of the blouse.

Recognising and Appreciating Resources

Recognising all resources makes it possible to free one resource for more satisfying use by substituting others. For example a family has sufficient money to buy its clothing as readymades. However, the housewife has enough time or the ability to make many of the clothes. Further more, she enjoys the creative aspects of sewing. Thus she frees money for a family trip which give satisfaction to all the family members.

Expanding Appreciation

This is a more subtle process that increases utility. It results in growth of the individual using the resource. For example the use of time to get a book from the public library and then read it may open doors to new and growing interests. To expand appreciation, combination of resources is needed like using one of the intangible resources such as knowledge or attitude along with one or more of the more recognised ones like money or time.

CONSERVATION OF RESOURCES WITH SPECIAL REFERENCE TO SHARED RESOURCE

We have already seen that the resources are limited and need to be managed. Now let us discuss how can we conserve these resources to ensure their availability for a long time.

Resource Consciousness

It means being aware and alert about resources. Sometimes we are not conscious about our resources, like human resource, which include the talents, skills etc. of the family members and shared resources like pure drinking water, clean and pure air, green cover etc. Being conscious about all the resources directs consciously and unconsciously our activities towards their conservation.

Adequate Supply

Ensuring adequate supply of resources also helps in conservation of resources. This is true specially in the case of larger family with lesser income. Such a family should make adequate efforts to supplement their income in a number of ways. Similar efforts should be made in the cases of other resources, especially shared ones which are in short supply like fuels for cooking, land for housing etc.

Use More of Abundant Resource to Conserve Scarce Resources

The resources, which are available in abundance should be used more than the scarce resources. For example a family which has many adult members and thus lot of energy available but has less income, should distribute work among its members by using their human resources instead of hiring domestic paid help to do their household work. Also a family can use solar energy, which is free of cost, available in abundance for cooking rather than spending non-renewable and scarce fuels like coal, wood or LPG, and saving money simultaneously.

Avoiding Wastage

Resources help in reaching goals, therefore, should never be wasted. A homemaker can use her extra time, energy and talents for productive work rather than waste it in idle gossip. Another very important example is avoiding the wastage of drinking water or electricity, which can be done in a number of ways. Close the taps properly when not in use, fill the bucket before bathing rather than using shower water to avoid wastage. Switching off the lights and fans of the rooms where no one is sitting, all the family members spending time together in one room so that the wastage of electricity can be avoided.

Economical Use of Resources

Whether money or human resources, a family should use all its resources economically. Usually people are careful regarding their money and material resources, which are tangible and more easily recognised but forget the same for human resource. The time and energy of family members, fuels for cooking and transport, portable water, electricity etc. should also be used economically, so that their utility increases.

Resource Sharing

Resources can also be conserved by sharing them. For example, car pooling or making use of public transport saves lot of petrol. Similarly work can also be shared. In a joint family, where there are more number of people, can share their household work thereby saving a lot of resources.

Managing Resources Wisely

We have already discussed that resources can be managed and wise management can help in conservation of resources to a large extent. In the present times of resource crisis and shortage, applying managerial process in the use of every resource is very important.

Development and Use of Technology

Development of such technology, which spares scarce resource by using less of scarce ones, can help in conserving resources. For example, developing technology to enable extensive use of solar energy, biogas and wind energy as domestic fuel or, for heating water or even for running vehicles will enable its use for these purposes possible. At present many of the technical advances are based on resource conservation, and the families should also make efforts towards this goal.

Recycling and Reuse of Resources

Recycling of waste and reuse of certain resources is possible. Reusing paper can help in conserving forest by reducing the demand of wood for paper. Many of the household wastes like water, garbage etc can easily be re-cycled and re-used for purposes like watering garden plants, biogas etc.

Making Wise Choices

Making choices or taking wise decisions regarding the use of the resources available can also help in conserving them. This factor has been well discussed under the head, maximizing satisfaction in the use of resources.

Water Harvesting

Rain water can be collected and stored in underground or roof top storage tanks to be used in times of scarcity. A part of it can be allowed to percolate into the ground to raise the level of underground water. This process of collecting rain water and conserving it is called rain water harvesting. This can be practiced both at household and community level.

Composting

The garbage from garden is put into a pit in one corner of the garden. At the end of each day, it is covered with ash and leaves. Gradually the lower layers are converted into compost or manure. This manure can be used for gardening and will conserve land resource.

Use of Alternative Sources of Energy

Renewable and pollution free sources of energy should replace the conventional and non-renewable ones. The following are the few such sources:

Biogas

It is a product of fermentation of animal manure in the absence of air. It chiefly consists of methane gas which can safely be used as a fuel for cooking as well as lighting. Ordinarily, a small biogas plant fed by the manure of 2-3 animals can produce enough gas for the daily cooking and lighting needs of a family of four persons. In addition, it can also be used to pump water or run small motors of lesser horsepower. Bigger biogas plant can be initiated and maintained also at community level.

Solar Energy

Solar energy can be used with the help of solar cookers, solar cells for lighting and solar water heaters. These newer technologies help in tapping abundant solar energy, especially in tropical countries like ours.

Hydel Energy

Water falling from height is used for generating electricity. Hydroelectric projects like Bhakra-Nangal project or Damodar Valley corporation (DVC) use hydel power for generating electricity at a national level in our country. However, small quantity of water falling from great height can also be used for the purpose. This is especially useful to set up local electricity supply in the far flung geographical areas which are not covered by national grid system. Thus, this energy source can be used at a community level to fulfill the electric needs of that community.

Wind Energy

Wind energy has been used since long time now to sail boats on water and to grind grain by setting up wind mills. Now with the technological advances it can be used for generating electricity as well. Low cost wind machines can be used at individual as well as community level to provide electricity.

However, community participation is essential for the management of these shared resources for their maximum utilisation as well as for their conservation.

Thus it is clear that a homemaker who wish to maximize satisfaction can do so by applying her managerial skills. A house where the managerial process is strictly applied is likely to utilize its resources better, thereby getting more satisfaction as compared to a family which turns a blind eye to its better management. On the whole, expanding the availability of resources a change in attitude of the family members, substituting one resource for another, recognizing the availability of resources, increasing their utility and expanding appreciation of the available resources together with better management skills will help a family to maximize their satisfaction level. In the end, every family should recognise, economize and remember the **three "Rs"** of resources — **Reduce**, **Re-use** and **Re-cycle** for sustaining the availability of resources to all.

7

Time Management ◀◀◀◀←

TIME management has been a significant aspect of the management of all resources. It helps to determine the worker's timings of a standard day's work, and to define the amount of time needed for a particular work. Time management is required for all types of workers irrespective of their nature–manual workers, home based workers, unpaid workers, skilled workers, semi-professionals and professionals.

Time, being an intangible resource, is always taken as the one that can easily be managed. Time is the only resource which is equally distributed to all persons. Everyone has the same amount of 24 hours of time in a day. Both the rich as well as the poor are alike as far as time is concerned. But, there is more emphasis on time management practices in today's context because, with changing life styles, urbanisation and women taking up gainful employment outside the home, there is greater need for optional, efficient and effective use of time.

Time provides an organising media for our lives a common denominator— within which we operate. Ongoing flow of time is an universal experience. What and how we make use of it is a subjective phenomenon. Hence, control of time requires a reconfiguration of the past, present and future— past as the basis for the present happenings, and present as the framework for the future. The idea of time is formulated from experience of changes in ourselves and around us. Appreciation of consequences of living in time provides a framework for reference.

Developing a time sense is a help in managing time. People differ in their ability to gauge the passage of time. The individual who has a good time sense has a valuable aid in controlling the use of time. Accepting time as manageable resource can add order to life.

The allocation of time has a major effect on the quality of one's life. The knowledge as to how such time is spent by each person, for work, individually or in group and how much time to be spent on non-productive work designated under rest and recreation is of value to the understanding of the life-style and the resultant quality of life of an individual.

NATURE OF TIME

Among all the resources available to us, time as a resource, is one of the easiest to measure, but one of the difficult to manage. Whether time is a human or non-human resource or is a resource at all, is still a factor that is being debated often. In all, time is still regarded as a resource, because it is used for all practical purposes. Similar to the material or energy resources, time resource can also be successfully handled; we can earn it, and spend it.

The abstract nature of time, individual perception and orientation to it has always attracted the attention of many thinkers and professionals alike.

Of all the resources, time is one of the easiest to measure, but one of the most difficult to understand. The understanding of nature of time requires an insight into human psychology and human behaviour. In the psychological literature, it is cited that *time is an act of mind, of reasons, of perception, of intuition, of sense, of memory, of will, etc.*

The continuing and compelling experience of time can be felt in our daily lives. The passage of time can be experienced through changes in events/reasons, process of growth and aging in all living-beings.

Everyone, irrespective of the position one holds in society, has the same 24 hours which one has to spend and it is also true that time once spent cannot be recovered. It is irretrievable. One can only invest it wisely and get the maximum utilization out of it. And, therefore, how one utilizes his or her time becomes important.

Perception of time varies in a number of ways. For most people, it is the *clock time*, which stems from the regular movements of the earth in relation to the time. Though it is the most accepted measurement of time, still scientists are making efforts to find out a more accurate and more easily measured standard unit for time. *Biological time*, or the cyclical occurrence of certain bodily symptoms has been felt by many, and has been noted in medical literature for long. The biological symptoms of time passage can be observed in certain situations of our day-to-day life like feeling hungry at meal times, sleeping patterns, and toilet habits; getting up at the same time in the mornings, i.e, waking up early even on a holiday. This kind of symptoms are felt more by the people who travel often across the globe and express it as *jet lag*. Another important aspect of *biological time* is that it helps a person observe, understand and self-assess the time of the day when one's potential is at its peak and that is the most productive time of an individual. For example, some students feel mentally agile and active in the morning and their concentration levels are high and hence, study is more productive and beneficial. For others, the most productive time can be evenings or *late-nights*. Similarly, some *housewive's* may have their most productive time in the mornings when they are able to accomplish more number of tasks. For other's the productive period may be towards evening or at night.

There is a kind of sense regarding the passage of time and can be termed as *psychological time* which a layman can easily understand. This is an inborn time sense and is a kind of judgement developed through practice especially in early life. Another aspect of psychological time is the sense of it, passing quickly or slowly. This kind of feeling comes in part from the quantity and importance of what is to be done in a certain amount of *clock time*. For a well-prepared student, writing answers within a specific period, time will seem to pass quickly or in short supply. While for a person waiting for a friend at a bus stop, time will pass slowly or will seem long. Whatever may be the differences in the perception of time, for all those who respect work and value punctuality, time is as much worth as money!

BALANCE IN TIME USE

An important guide to the use of all resources of time is "balance". Balance in the use of time has three classic divisions—work, rest and leisure, with some approval in the folkways of an equal division among the three. Balance of time means that there must be time for work, rest and sleep, with sufficient leisure for some phases of living that will keep a person emotionally stable and intellectually alert.

Leisure is a free time after the practical necessities of life have been attended to. Arithmatically, leisure means the time left after work, sleep and other necessities. But, the actual meaning of leisure differs from the arithmetical meaning. Actually leisure is the time which one uses as he pleases. One chooses to use one's leisure hours without compulsion. The tone of any society is to an appreciable extent determined by the quantity and quality of the leisure of its members and by the way in which it is distributed in the society.

Leisure is a by-product of mechanisation for which both the individual and the society may be unprepared. Workers do not know how to live with work and leisure. Leisure comes out in the open confronting its creators with baffling problems, as to how it should be utilised. Formerly it was a privilege of a small upper class. Now this unplanned gift comes to many who are unprepared and lack the back-ground for making use of it.

Work is an important part of the average person's day. Therefore, attitudes towards work may be distinguished. Some people have the feeling that work is a duty, and is so important that its accomplishment must overbalance everything else. A time plan based on this philosophy will revolve around work, other things fitted in as remaining time permits. Some people look upon work as a necessary evil and will plan to get it over as quickly as possible, either by doing less of it, doing it less well, or occasionally doing it more effectively. A third attitude is that of looking upon work as a creative experience. Not every job has equally creative possibilities, but some of the varieties of work done in an ordinary home may be very stimulating in themselves.

Besides facing one's own philosophy of time management, knowledge of time patterns of various household activities is an aid in effective management. In addition, other tools like the handling of peak loads, taking advantage of the implications of the work curve and applying work simplification principles add to the effective management of time.

There are various factors that determine the time use pattern of the homemakers. The most important of these is the stage of the family life-cycle. During the beginning and the establishment stage, the time demands are much greater than in the contracting stage. Also, the type of the family, i.e., nuclear or joint, affects the time use pattern of the homemakers. In Indian homes, members of the joint family equally participate in the household chores, and the demands on the time of the homemaker, are lessened. Besides, size and composition of the family also affects the use of time. If there are more number of young children in the family, then the time input by the homemakers increases, while in a family with more adult members, the work is shared and homemakers need to spend less time on household activities. If there are more older people or male members in the Indian family, then also the time demands of the homemakers are increased.

Time Patterns and Time Costs

Thus we find that different people have different kinds of demands on their time use. However, everyone learns to use their time effectively by following some kind of time patterns.

What guides can home-makers use in planning and evaluating their expenditure of time? How do other home-makers use their time? Which home-making activities are most time consuming? How much and what kind of help do other home-makers have ? How do other home-makers spend their leisure time? How do home-makers who ae regularly employed outside their homes meet their home-making responsibilities ? Some of these questions will be answered by studies of home-maker's uses of time.

These studies agree in general on the rank order of time spent on groups of household activities and on the approximate time spent on each group. Also, total weekly times are fairly in accord, ranging from 47 to 52 hours. In the 1950's two careful studies were published by Weigand in New York and Cowles in Wisconsin . Both studies verified in general the earlier findings and added to the detailed knowledge of time patterns.

The general findings are that food activities take approximately twice as much time of the home-maker as any other group of activities out of which, clearing away requiring roughly one third of the food time and preparing taking two -thirds. House care and the clothing and textile activities rank similar in time demands upon the home-maker and are second in importance .The other two classes of activities, care of family and purchasing and management, trail the first three groups. Either or both of them assume greater importance as food and, to a larger extent, clothing activities absorb less time. (Table 7.1).

TABLE 7.1: FULL TIME AND EMPLOYED CITY HOME-MAKERS' USE OF TIME FOR HOMEMAKING ON ONE WEEKDAY*

| Homemaking Activities | Type of Homemaker | | | |
| | Full Time | | Employed | |
	Hours	Per cent of Homemaking Time	Hours	Per cent of Homemaking Time
Food preparation	1.6	22	1.2	29
Dishwashing	1.0	13	0.7	17
Care of the house	1.6	22	0.8	20
Care of clothes	1.6	22	0.8	20
Care of family members	1.1	15	0.3	7
Marketing, records	0.5	6	0.3	7
All home-making	7.4	100	4.1	100

*From Elizabeth Wiegand (July, 1954). Use of time by Full-time and Part-time homemakers in relation to home management. *Corwell Agr. Exp Sta. Memoir*, **330**.

A recent study was conducted by Batra S. and Seetharaman P[1]. on *Perception and utilisation pattern of time among urban women in different occupations (2000)*. For the purpose of calculating time norms, the activities performed by all home-makers were classified in four categories as follows:

- **Home-making Activities**
  Preparation of breakfast, lunch, snacks and tea, dinner, care of clothes, house cleaning, care of children, Religious Activities, Entertaining guests, gardening.

- **Employment Related Activities**
  Employment related activities carried out at work place and at home, transportation to and from place of work.

---

[1]Batra, S. and Seetharaman, P. (2000). *Perception and Utilization Pattern of Time among Urban Women in Different Occupations.* unpublished doctoral thesis submitted to the University of Delhi, Delhi.

- **Personal Activities**
  Personal care, sleep and relaxation , leisure

- **Miscellaneous Activities**
  Marketing, General Reading etc.

Time Norms for Household Tasks

Few attempts have been made to establish norms or average times spent on specific household tasks such as bed- making, washing windows and so forth. A few scattered ones are known, for example, the norm for ironing a man's shirt by the best method seems to be about 6 minutes or less, for making a bed 7-8 minutes. Whether or not established norms exist and are known to the home-maker who wishes to manage her time, a knowledge of her own time costs is extremely important. Over a period of years a home-maker probably knows about how long it takes her to complete most of her various repetitive tasks. An inexperienced home-maker can quickly find out her norms by recording for a few days the amount of time she uses on each job. The averages of these individual records become her personal time norms.

The study by Batra S. and Seetharaman P. (2000) also reported Time norms for various activities. These are as follows :

- **Homemaking Activities**
  The full time home-makers spent maximum time, 6.58 hours on homemaking activities followed by business class, 3.92 hours. The least amount of time was spent by service class, i.e. 3.05 hours on this activity in a day.

- **Employment Related Activities**
  The respondents from service class spent maximum amount of time, 6.63 hours on this activity followed by business class 6.54 hours and professional class 6.46 hours. The least amount of time, 5.04 hours was spent by academicians on employment related activities in day.

- **Personal Activities**
  The respondents from Academic Class spent maximum amount of time 10.22 hours on personal activities followed by full time home-makers, 9.51 hours. The least amount of time, 8.26 hours was spent by professional class on personal activities in a day.

- **Miscellaneous Activities**
  The full time home-makers spent maximum amount of time, 3.19 hours on miscellaneous activities followed by professional class, 2.46 hours and business class, 2.45 hours. The least amount of time, i.e. 2.21 hours was spent by service class on miscellaneous activities in a day.

Norms for Leisure Time

Some norms are available for leisure time activities of home-makers. Leisure activities may be broadly interpreted as anything not classified as work or rest, and in this interpretation include such uses of time as eating and dressing. They may also be more rigidly interpreted. There is evidence that even in the most restricted sense, leisure is a fairly important part of a home-maker's day. The Weigand study shows a daily average for full time home-makers of approximately 4 to 5 hours spent on community activities and other leisure time activities.

Sorokin's study of use of time of 22 men and 81 women, relief workers under the Works Progress Administration, offers a breakdown of leisure activities separate from physiological and economic needs. Some of his terms of classification are self evident. Others are used in specialised ways. Societal activities emphasise interaction with others and also performance of altruistic acts. Pleasurable activities are miscellaneous activities performed with or without others, such as auto riding, taking part in physical sports, and idling. The pattern of time use is as follows:

TABLE 7.2: AVERAGE TIME SPENT DAILY PER MEMBER OF THE TOTAL GROUP

Activity	Hours	Percent
Societal	1.35	27.2
Religious	.14	2.8
Intellectual	1.14	28.4
Artistic	.41	8.2
Love and courting	.15	3.0
Pleasurable	1.51	30.4
Total	4.97	100.0

The total of nearly 5 hours per day falls within the range given above for non work activities of home-makers.

Work Units

In studying patterns of use of time in order to establish time norms of household tasks, there has been some attempt made to go beyond mere statement of what activity is taking place. A pioneer effort to measure how much of the activity was accomplished in the time used for it was made in the late 30's by Jean Warren in her development of work units and work loads. The purpose of the work unit was to enable the amount of time required in one household to be compared with that used in other households accomplishing the same quantities for the same tasks. It must be recognised that the work unit and work load resulted from quantitative analysis, and neither in any way indicated the quality of the work accomplished nor the equipment used. A unit for qualitative as well as quantitative comparison of household activities would be a step further in providing norms for home-makers.

Through later research at Cornell carried out by Kathryn Walker, utilising Wiegand's time study, six types of quantitative work units were developed. Although the concept is universally applicable, these units were evolved in New York state and may or may not be usable in different geographical locations. They will also need revision from time to time as homemaking practices and conditions change.

Walker's definition of a work unit in homemaking is *the amount of household work done in one hour under average condition by an average worker.* The work load is the sum of work units. Her six types of work units are: dishwashing (related to number of persons in the household), meal preparation (4 types of meals), physical care of children (related to age), washing clothes (by tubful), ironing clothes (by number of pieces), and regular weekly care of house (with and without children 18 years of age). Thus in setting up the units, important factors determined the number for each situation, as shown in the example below of work units for care of household (Table 7.3)

TABLE 7.3: WORK UNITS FOR CARE OF HOUSEHOLD

Tasks	Composition of Household	Work Units
Regular care of the house (All daily and weekly tasks)	All adult household Household with children under 18 years of age.	4.0 per week 7.0 per week

TOOLS IN TIME MANAGEMENT

There are various tools involved in the management of work time. It is very important to consider them while managing time. They provide the basis for time management. These are as follows:

- Peak Loads
- Work Curves
- Rest periods, and
- Work Simplification

Peak Loads

This is one of the important tools to be considered while managing work time. For most people activities pile up on each other at certain times of the day or the week or the month or the season. These packed periods are called peak loads. For example, for a homemaker peak loads can be daily, weekly or seasonal such as the time of breakfast and getting the family off for the day is a daily peak load, the thorough cleaning of the house a periodic peak load and the Diwali or a festival preparation, a seasonal peak load.

The wise manager of time will, if possible, level off peak loads either by starting the piece of work early enough to avoid the last minute rush, or by completing some regular work in advance so that more time may be made available for reducing higher demands at the peak loads. In some instances, the peak load may be lessened by delegating some work to other family members or by adopting some work simplification methods. For example, the housewife can assign some duties to her daughter or to any other member of the family to reduce her daily peak load or use a convenience food in the breakfast menu, especially in the morning. Awareness of the peak load and the methods of handling them is an important tool for managing time. This is even more important for a homemaker who is gainfully employed outside the home.

Work Curves

The second tool for managing work time is the work curve. This is a device originated in industry to study the changes in output of work over a period of time. When a worker is engaged in a task which is accomplished in identical units, like, the number of sheets ironed per hour, the quantity of those units produced in a given amount of time may be taken as an index of his accomplishment. It has been found that the rate of production follows a some-what typical form-typical of the work day or to a less extent, of the work week. A typical work curve has the following features:

- Starts sluggishly
- Sharp rise as worker gets into stride
- Falling off in the middle of the spell with a fresh spurt as work nears its end
- Find falling at the last hour

In analysing a hypothetical work curve shown in the figure 7.1 with a lunch break, the preliminary increase, 'a-b' signifies the warming up period known as the 'warming up'. The letter 'b-c' indicate the plateau of greatest steady production, 'c - d' shows the first major drop in production. The beneficial effect of the lunch period break is shown in the figure. Production starts out at a higher level after lunch than in the morning but never reaches as high as 'b-c' in the morning. The figure shows a decrease in production and that is due to the effect of accumulated fatigue at the end of the day. The worker may stop at 'f' or continue to 'g,' but the production level will decrease from 'f' onwards. The drop from 'c - d' is supposedly due to boredom developing during the job if the work is light. In heavy manual labour the final decrease will probably be very great and it is possible that output may fall to zero if work is continued further to the point of exhaustion.

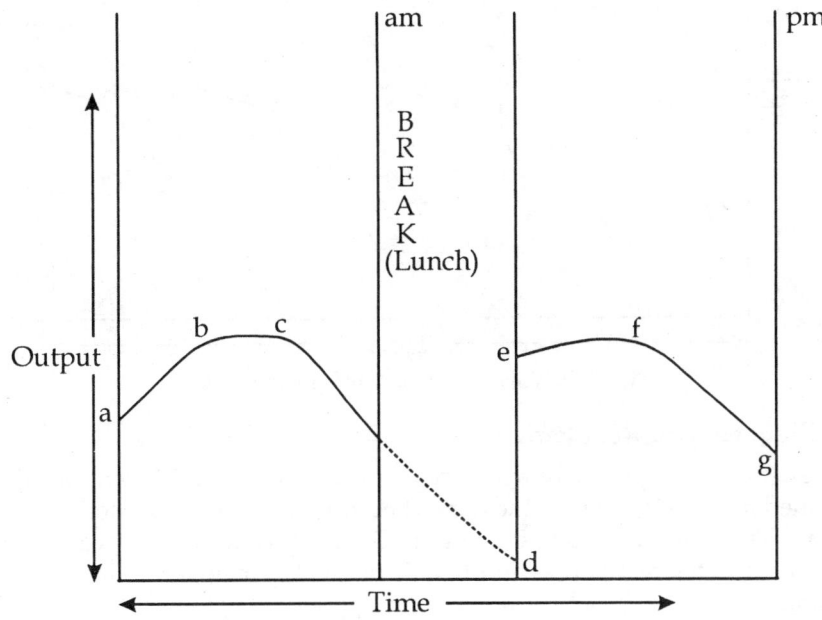

Fig. 7.1: A Typical Work Curve for a Day

A practical and valuable application of the 'warming up' by the homemaker is doing as much as possible at one sort of task before changing the type of work. In cleaning a group of rooms, for example, the woman who does all the dusting at one time, or all the vacuum cleaning or all the window washing in one attempt is doing more of each job on the plateau of greatest output than the women who jumps from job to job and back again.

A most desirable work curve is done in which ab warming up (wu) is (a) steep line showing that worker got into the swing of work rapidly and (b) achieved a high plateau of production - longer the plateau of production, greater the output of work accomplished.

Rest Periods

This is the third tool for time management. All of you must have observed the favourable influence of the rest period on your output of work. A rest period need not mean complete cessation from work, although that is desirable after a heavy manual labour. The greatest results can be expected if the worker lies down and relaxes completely, because reclining requires less expenditure of energy than any other body position. As compared to most workers, the homemaker can provide good conditions for rest more easily. However, if it is impossible to lie down, complete relaxation in a sitting position will relieve fatigue. The completeness of relaxation during a rest period probably determines its success. A change in the type of work may also serve as rest periods for each other. A sitting down job after a walking one, or a mental task after one requiring physical exertion will also prove restful. Now the question is how long a rest period be and after what interval is it required?

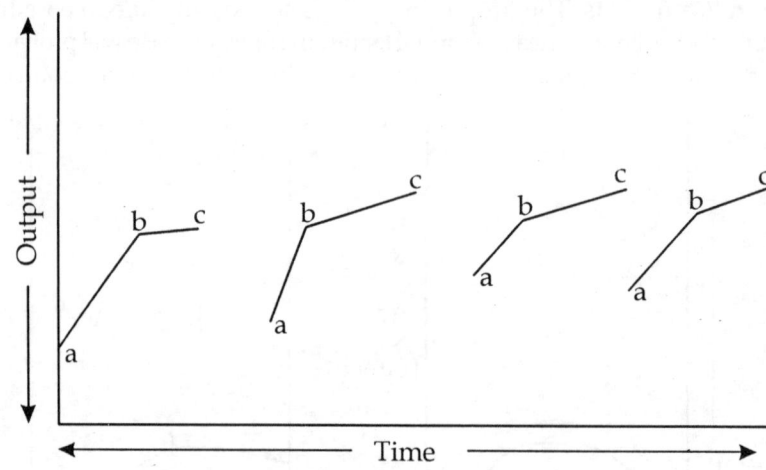

Fig. 7.2: Work Curve of an Erratic Worker

Length and Frequency of Rest Periods

The optimal length and frequency of rest periods in relation to increases in production is more easily determined in industry, than in the home, because the homemaker does a variety of jobs in a single day and has no accurate record of output. Often it is difficult to convince the homemaker of the effectiveness of the rest period and here it must be remembered that the results are not always the same.

The homemaker should have no feeling of guilt when she lies down for a few minutes during the busy part of the day. The homemaker must determine for herself the frequency, length and type of rest periods which she considers as most desirable. She should recognise that a change of work may rest her if the present task is boring, or if a particular set of muscles is tired. For example, a homemaker stitching clothes on sewing machine can shift to put buttons or doing hemming in between or just relax her leg muscles and then again continue with her stiching when the hand muscles get fatigued.

In the case of physiological fatigue, she must realize that longer the rest period is postponed, the longer it will take to recover. She must learn to relax thoroughly during rest periods in a cool, dark and quiet place, if complete rest is desirable. Thus the length and frequency of rest periods depend upon the kind of work as well as how long the work has been done. It should be introduced when the output slows down, but it should not be too long to lose the benefits of *warming up* gained earlier.

Work Simplification

It is also one of the tools of time management which is related to energy management, as it includes improving methods of work which requires lowering both time and energy expenditures, because the time and energy are required to do any task and they largely depend on the hand and body motions used.

In simple terms, work simplification is the conscious seeking of the simplest, easiest, and quickest method of doing work. Or in other words we can say that work simplification means improvement in performance of task. Improvement in performance of a task means that the work is made easier because the new method is a more convenient one, permitting smooth, natural and rhythmical motions.

It is a very important tool of time management which not only is helpful in an industry but a very useful tool for a homemaker who wants to have a liberal supply of time as well as the need to conserve her energy . This tool of time management well be discussed in detail later in the chapter on energy management.

Value of and Bases for Time Plans

Managing time invovles both making plans and carrying them out. A time plan shows what one expects to do in a given period such as a morning, an afternoon, or possibly during a whole day or a week.

Most home-makers follow some sort of time plans. As might be expected, all do not manage time equally well. Some studies have shown that home managers tend to fall into three groups :

1. a small group for whom home work is a minor consideration
2. a large group who by careful management fit most of their work into the morning and into an hour or two in the afternoon with many afternoon and evenings relatively free for the children, their husbands, social life and community activities.
3. a large group for whom each day is a *nip-and-tuck* race to accomplish the things that must be done between morning and bed time with occasional afternoons and evenings free only by planning for them well in advance.

The success or failure of these groups of homemakers is due largely to their approach to homemaking responsibilities. Homemaking and managerial skills can be learned if one is interested and willing to work and study.

Value of Time and Activity Plans

Part of the usefulness of a time plan lies in the fact that it leads one to think through many work problems in advance. Thinking these problems thoroughly does away with indecisions and free's one's mind for solving other problems and also for meeting constantly arising new situations. Learning to make and carry out daily time plans gradually establish thought patterns that flow along automatically. Getting breakfast soon becomes a routine task during which one can make a dozen-and-one decisions that most families must make at the start of the day. Such automatic methods of work promote speed and efficiency and free time for other responsibilities. Carefully thought-out time and activity plans are useful tools, for they save time and energy and relieve the tensions created by indecision and uncertainty.

Bases for Time Plans

To build an efficient and workable time plan, each homemaker , with the help of her family , must decide which activities ought to be carried out each day, which during the week, which are the most important, the ones that are in any case must be left out and what special and seasonal tasks should be fitted into daily and weekly plans. They also need to think through the problem of the time in which they can best do each task. Although many daily and weekly tasks remain much the same throughout the year, special and seasonal tasks and recreational activities constantly keep changing. Each of these makes different time and energy demands and may bring about other changes in plans.

In most homes the time for doing certain tasks is definitely fixed by some condition outside the home or some special demand within it. The husband's or children's hours of work or the children's hours at school usually determine the time, breakfast must be prepared and served, when lunches must be made, when children must be taken to school or others to work. Although the time at which certain tasks must be done is relatively fixed, we do not have to adhere to the same rigid schedule in the performance of others.

Each member of the family may also have individual tasks or activities that must be done at a definite time. Some of these may last only a short time, others, a long time. Music lessons, scout work, school and outside activities all make, definite time demands. These activities must be considered because they affect both the use of the homemaker's time and the amount of help other members of the family can give in the house.

The distribution and sequence of other homemaking activities will depend, of course on when the fixed tasks must be done and how much time they will take. Definite time tasks shape daily and weekly time and activity patterns in all homes .

The time when help is available also frequently determines when certain tasks can best be done. Jobs that require several workers must often be filled into the plan at a time when the workers are free. When outside help is employed the time at which the help can work also fixes the time at which many paid jobs are to be done.

Time and Activity Patterns

The daily and weekly time and activity pattern in all homes reflects the interests, work habits and personal needs of the family. For instance, weekends may make heavy demands on time and energy for many homemakers, so they may find the work load easier to carry if they wash and iron on Tuesday, or Tuesday and Wednesday, reserving Monday for less heavy jobs. Others interested in outside activities, may prefer to wash and iron on Monday, or Monday and Tuesday, so that they can free large blocks of time on the two consecutive days later in the week. Many who do not have enough energy to carry the entire homemaking load have the laundry done outside the home or use more prepared or convenience foods than they otherwise would.

In most homes the distribution of time for recreation varies from week to week. At times, planning for some recreation every day may be desirable. At other times it may be advantageous to omit most of the daily recreation for several days in order to have a long period for time demanding activities such as an all day meeting away from home, a movie, a game of golf , a picnic, or a week end trip.

The work habits and the type of work being done influence the amount of rest the homemaker needs during the day. Short rest periods may seem the best for some, and a change of work may give others enough rest and relaxation.

FACTORS TO CONSIDER IN MAKING TIME AND ACTIVITY PLANS
Peak Loads

Work and activities tend to pile up in most homes at certain periods. These peak load periods may occur at certain times of the day, week, month or season. For some homemakers the daily peak load may come when they are preparing dinner and getting the children off to bed , the weekly peak load, doing the washing and ironing , the monthly peak load, cleaning the windows ,and the seasonal peak load, preserving food.

In all homes there is less danger of periods of overloading if the homemaker plans to distribute work and activities over the week, so that the demands on time and energy are about the same each day. Alternating heavy and more fatiguing jobs with lighter, less fatiguing ones, and planning to do only comfortably heavy work at one time help distribute the energy cost expended during the day.

A yearly calendar of all special and seasonal activities , vacations, holidays and anniversaries is a great help. It enables the homemaker to look ahead and see when and where she can and must fit these special activities into current plans. Long range planning of this kind enables one to keep tasks from pilling up and eliminate the nervous strain and fatigue that come from carrying an excessive load and from working under pressure.

Sequence of Activities

The sequence or orders in which activities follow one another in the day's schedule is in part determined by the definite time tasks, in part by the relationships of tasks to each other, and in part by ourselves. All activities should be arranged so that they can be carried out with the least amount of effort and tension. Time and motions can be saved by grouping activities that normally follow one another in a sequence and that fit into a time block and by allowing some activities to overlap. Overlapping activities are at least two activities that are going on at the same time, one is active, the other inactive in the sense that the worker only needs to keep her mind on it, or that it may require only occasional attention, like baking a pie or cake while ironing or washing windows. By combining tasks that require the use of the same tools, one can reduce the time and energy needed for assembling equipment and materials and for cleaning up and putting them away. Each one must determine the best order for her own situation. By studying one's own work habits and trying different plans it is usually possible to find the one that saves the most walking and interruption.

Time Required for Different Activities

Estimating the amount of time for each activity or group of activities will probably take, is important in making a workable plan. Without this information arranging a closely knit plan or deciding how much work one can comfortably do each day is difficult. A knowledge on the time norms for various activities will be of great help in determining the time required for these activities.

Time for Emergencies

We sometimes find weekly plans difficult to follow because of unplanned and unexpected interruptions or demands that have to be met. *Out-of-town* guests may drop in and spend the night, an aching tooth may require several visits to the dentist, a bad cold may keep one of the children in bed, a clogged sink may delay dishwashing, the appearance of moths in a closet may mean extra cleaning and spraying, a costume for a school play may have to be made, a special meeting may be called, and so on, are all unexpected time-taking and work making demands.

Such emergencies can best be met by allowing some free time in daily plans. *To meet emergencies in the family, it requires flexibility in personal time patterns, shift in management plans, and a resilience to recover when the trying hours are done.*

Steps in Making Daily and Weekly Time and Activity Plans

Once it is decided to make a time plan, next is as how to make it? Let us now see the various steps involved in making time plans.

A workable time and activity plan must be built step by step to fit the needs of one's family. Conditions in no two homes are the same. Plans for a family with young children will differ greatly from a family with teenage children. Some homemakers, such as doctors', and farmers' wives, must dovetail their work with that of their husbands and plan time for interruptions and unexpected demands, because they are a part of the day's activity. Although the details of time and activity plans will differ in each household, the steps in making them are much the same.

Step 1: Consists of listing the everyday, weekly, special and recreational activities of the family. The list will probably include such tasks as the following:

EVERYDAY JOBS AND ACTIVITIES
- Caring for children and invalids
- planning, preparing and serving meals
- Packing lunches
- Baking cookies, cakes, breads
- Preparing baby's food and caring for baby
- Washing dishes
- Doing unexpected tasks
- Making beds
- Caring for house
- Caring for pets
- Resting and personal care
- Recreation and social activities
- Doing farm work
- Doing any other tasks.

WEEKLY AND SPECIAL TASKS AND ACTIVITIES
- Washing and ironing
- Mending and sewing
- Cleaning house thoroughly
- Other special cleaning

- Shopping and ordering
- Washing Windows
- Special cooking and baking
- Preparing special meals
- Going to meetings, clubs etc.
- Doing any other tasks
- Going to doctor, dentist etc.
- Errands away from home
- Doing farm tasks
- Making repairs on equipment and house
- Going to bank and keeping accounts
- Engaging in recreational activities.

SEASONAL TASKS AND ACTIVITIES

- Planning and directing children's vacation activities.
- Preparing for holidays
- Preparing for birthdays
- Storing seasonal clothing
- Preserving food
- Sewing
- Vegetable and flower gardening
- Caring for yard
- Putting on and taking off screens and storm windows
- Doing any other tasks.

Step 2: Consists of making a plan for everyday or routine tasks and starting or underlining those that must be done at a definite time. This provides a skeleton around which build the rest of the plans. Such duties as preparing and serving meals, packing up lunches, taking children to school and picking them up and daily cleaning are included in this skeleton plan. After the daily or routine tasks are allocated and timed, blocks of free time will be left in the morning and afternoon for weekly tasks and activities.

Step 3: Consists of completing the weekly plan. At this point we must fit the weekly, special and seasonal jobs into the free blocks of time in the daily plan. In allocating these jobs the homemaker must consider the needs of her own household, the work habits and free time of members of the family.

Step 4: Consists of deciding who will do various tasks and calls for group discussion and planning. In doing this, the work carried out by the mother and father individually as well as the responsibilities they share together and the duties of each child , no matter how small, are clearly defined and understood. Step 4 is usually combined with steps 2 and 3, because many of these decisions are made when the order and time of work are being determined .

SAMPLE TIME SCHEDULES

As it is mentioned earlier, it is necessary for a homemaker to prepare time schedules .Though it is difficult to make a common time schedule for all homemakers, a sample time schedule is indicated in Table 7.4 to 7.8. They include a Daily, Weekly and Seasonal plans for a homemaker, who is not

TABLE 7.4: DAILY PLAN FOR A NON-GAINFULLY EMPLOYED HOME-MAKER

From		To		Plan
5.30	a.m	6.00	a.m	Personal care like getting up, brushing, taking a cup of tea etc.
6.00	„	7.00	„	Prepare and pack tiffin for self, children and husband
7.00	„	7.45	„	Prepare and serve breakfast to the family members and self.
7.45	„	9.00	„	Clear table, supervise kitchen work as maid washes dishes, daily cleaning , straightening and bed making
9.00	„	10.00	„	Sorting and washing clothes by hand or machine, putting these clothes for drying etc., bathing and dressing.
(10.00	„	11.30)	„	Provision for weekly tasks and activities
11.30	„	12.00	„	Rest and reading newspapers, magazines, watching television etc.or some leisure time hobby work like sewing embroidery etc.
12.00	Noon	1.00	p.m	Prepare lunch and semi preparation and planning for evening snacks and dinner
1.00	p.m	1.30	„	Watch TV or do some hobby work/resting
1.30	„	1.45	„	Bringing children back from school/bus stop.
1.45	„	2.30	„	Serve lunch to children and cleaning up
2.30	„	3.30	„	Rest and getting in touch with friends and relatives on telephone, etc.
3.30	„	5.00	„	Help children with study and homework
(5.00	„	6.00)	„	Allow for weekly tasks and activities
6.00	„	6.30	„	Prepare, serve and have tea, snacks, with family members
6.30	„	7.30	„	Preparations for dinner
7.30	„	8.00	„	Help children pack bags, prepare for next day's activities.
8.00	„	8.30	„	Serve and have dinner with family members
8.30	„	9.00	„	Clear table, clean kitchen etc.
9.00	„	10.00	„	Spending time with family members/watching television etc.

gainfully employed during stage II of family life cycle with school going children. Among all the stages in a family life-cycle, the highest demand on the time is felt at stage II. Therefore the sample time plan focuses on her activities at this stage. As mentioned earlier, it is only a sample plan. Though many homemakers can follow this, they need to modify this slightly to suit their individual needs. A sample daily Time Plan is given in Table 7.4 for a non-gainfully employed homemaker. The weekly and seasonal time plan, as given for a non gainfully employed homemaker can also be followed by a gainfully employed homemaker with necessary modifications. Also, a sample time schedule for a college-going student is indicated in Tables 7.9 and 7.10.

TABLE 7.5: WEEKLY PLAN FOR A NON-GAINFULLY EMPLOYED HOME-MAKER*

(10.00-11.30 a.m.)

DAY/DATE	Plan
Monday	Attend computer classes
Tuesday	Mend and sew
Wednesday	Attend computer classes
Thursday	Interaction with friends , Recreation and social activities
Friday	Attend computer classes
Saturday	Shop for groceries and other items
Sunday	Spend time with family members at home or having an outing/visiting friends, relatives etc.

(5.30 -6.30 p.m.)

DAY/DATE	Plan
Monday	Drop children for tennis, music, dance or Hobby classes, and go for walk.
Tuesday	Same schedule is followed.
Wednesday	Same schedule is followed.
Thursday	Same schedule is followed.
Friday	Same schedule is followed.
Saturday	Family outing or spending time together on indoor games.
* Sunday	Spend time with family members at home or having an outing/visiting friends, relatives etc.

* On Sundays no fixed time schedule is followed, there is flexibility in time plans according to the needs, desires and requirements of the family and friends.

TABLE 7.6: SEASONAL PLAN FOR A NON-GAINFULLY EMPLOYED HOME-MAKER

SEASON	PLAN
Spring	Vegetable and flower gardening
Winter	Storing seasonal clothing and preparing the wardrobe for the next season.
Summer	Plan and prepare for holidays. Plan and direct activities for children during vacations. Preserving food - making pickles, chutneys etc., storing seasonal clothing
Monsoon	Vegetable and flower gardening

TABLE 7.7: A DAILY PLAN FOR A COLLEGE GOING STUDENT

From	To	Plan
6.00	6.30	Personal care like getting up, brushing etc.
6.30	7.00	Bed tea, read newspapers etc. and help mother in house work.
7.00	8.00	Have breakfast, get ready for college, do some work on the previous day's study.
8.00	9.00	Commuting to college
9.00	3.00	Attending college
3.00	4.00	Commuting back from college
4.00	5.00	Rest, Relax and personal care.
5.00	6.00	Interact with friends and socialise, miscellaneous activities
6.00	8.00	Study and miscellaneous reading and activities, help in housework
8.00	8.30	Dinner
8.30	10.30	Watching TV, Recreation etc. and spending time with family members

TABLE 7.8: WEEKLY PLAN (FOR A COLLEGE GOING STUDENT)

(5.00-6.00)

DAY/DATE	PLAN
Monday	Attending swimming classes (summer)
Tuesday	Play outdoor sports or attend a gym or Taekwando/karate classes.
Wednesday	Attend swimming classes (Summer). Play outdoor sports or attend a gym or Taekwando/karate classes.
Thursday	Play outdoor sports or attend a gym or Taekwando/karate classes.
Friday	Attend swimming classes (Summer). Play outdoor sports or attend a gym or Taekwando/karate classes.
Saturday	Play outdoor sports or attend a gym or Taekwando/karate classes.
Sunday	Flexible time schedule according to the needs and desires of the individual and the family, having an outing or a family gathering etc.

The time schedule as presented is for a normal working day of a college going student. It will vary during examination days and vacations where the utilisation of time will be different. Accordingly, the students should make time schedules depending upon their availability and demands of time.

Preparation of a Time Schedule

Since a time schedule requires more definite thought than a series of projects plan does, it should be carried out through a set of a fairly definite steps which are as follows:

1. List all Items to be Included, Grouping under Flexible and Inflexible

In some cases, there is only a fine line separating flexible from inflexible items. For example, going to classes is inflexible for a student, or feeding the baby is inflexible for a home-maker, whereas dusting one's bedroom or living room is a flexible task, which need not be done at a fixed time. It is equally important to break down lengthy or complicated tasks into parts. Not only are parts easier to grasp and to check on later, but they are easier to face and perform. Thus these activities are to be listed and arranged in a sequential order.

2. Set Down as Accurate an Estimate of Time for Each Task as is Obtainable

Settings down of time estimates may be accomplished in one of two ways: Either the person uses her own time norms found through long experience or through keeping a few records of repetitive tasks or she must estimate as best she can, the time required for each part of her schedule, get such information from other sources such as friends, relatives or books.

This time estimates are important, because the homemaker who takes two hours to clean a closet today should know that she cannot clean three closets in the same amount of time tomorrow or on any other day. Thus estimating the required time for each task is important ,prior to the preparation of a final time schedule.

3. Bring Total Estimated Time Needed and Total Available Time into Harmony

The third step in making a time schedule, bringing needs and wants into harmony, is the same process as that applied to the more tangible resource-money. This is done generally and felt more in the preparation of a budget in the financial management. This step calls for adjustment, and listing the earlier activity of flexible and inflexible plays an important role here.

4. Determine Time Sequence

It requires both listing jobs in order and determining logical times when they are to be done. Such as, build around the tasks that are fixed both as to necessity of performing them and/or clock time when they must be done, alternate light and heavy jobs as far as possible, include elasticity periods without fail. Sequence of activities can be planned most effectively, if we take into consideration the aspects such as fixed jobs, regular, routine, rest periods and the *warming up* period. This step requires a lot of care, since one has to bring a good balance in the use of time resource.

5. Write Out Plan

Now check the above sequence of activities before preparing the actual final plan of action. If the period planned for is short enough and soon enough, so that all the preceding steps may be accomplished while the plan can be remembered. Forms for writing out time plans vary from a separate card for a day with a few notations to a week's or even a month's plan. It all depends upon the convenience of the person for whom the plan is made.

6. Merge Individual Plans with Others for Co-ordination

The carrying out of the last step depends upon whether or not one is working alone. A homemaker may need especially to co-ordinate her plans with those of other family members, specially that of the adult members in the family, like her husband, grown up children, elderly parents etc.

After a plan is made, it should be tried out as carefully as possible and later evaluated. The time schedule can be altered slightly if there is a need for doing so. This may become necessary if some emergencies occur. Otherwise the time schedule can be periodically examined and re-organised to meet the changing situations and demands.

TIME DEMANDS DURING DIFFERENT STAGES OF THE FAMILY LIFE CYCLE

Although certain problems in time management are common to all stages of the family life cycle, the emphasis on each will vary from stage to stage in every family. Some of these problems are:

1. Recognising the importance of securing a suitable balance between work, rest and leisure time.
2. Considering all members of the family in making time and activity plans.
3. Taking time costs into consideration when making choices of goods and plans of actions.
4. Lowering time costs of homemaking activities.

Understanding the demands on the homemaker's time during the different stages of the family life cycle will help families plan ahead and prepare themselves to meet new and changing time demands.

Stage I: The beginning family is a period of adjustment and child bearing for young homemakers. At this stage, family goals, time and work patterns, work habits, and the division of responsibilities between husband and wife are established. Unless the homemaker is employed outside the home, demands on her time are light.

Stage II: The expanding family brings new and heavy demands on both the parents. The coming of children requires the greatest adjustment in the time patterns of homemakers. Farm and city homemakers with children under a year old used from twenty one to twenty five hours a week in caring for their families according to some studies. They used about thirteen hours when the youngest child was one to two years old. Thereafter their time demands decreased steadily until the youngest child was about nine years old. While children are in grade school and high school, their time demands, although different, remain high. They center around the problems of guiding and directing children in assuming their places as responsible members of the family, of making time plans together and helping each one to evaluate their use of time. As children approach and reach adulthood, demands on the homemakers time depend on whether the children go to college, marriage and leave home, or whether they take jobs and live at home.

Stage III: The contracting family covers the period during which the children have become independent. Mothers now have some free time to use as they wish. They can make the most of these years by learning to co-ordinate their time and recreation plans with those of their husbands. The curve in Figure 7.3 shows that the time demands upon the homemaker rise rapidly with the coming of children and remain high until the children complete their education and leave home. A gradual drop comes with the approach of the retirement period.

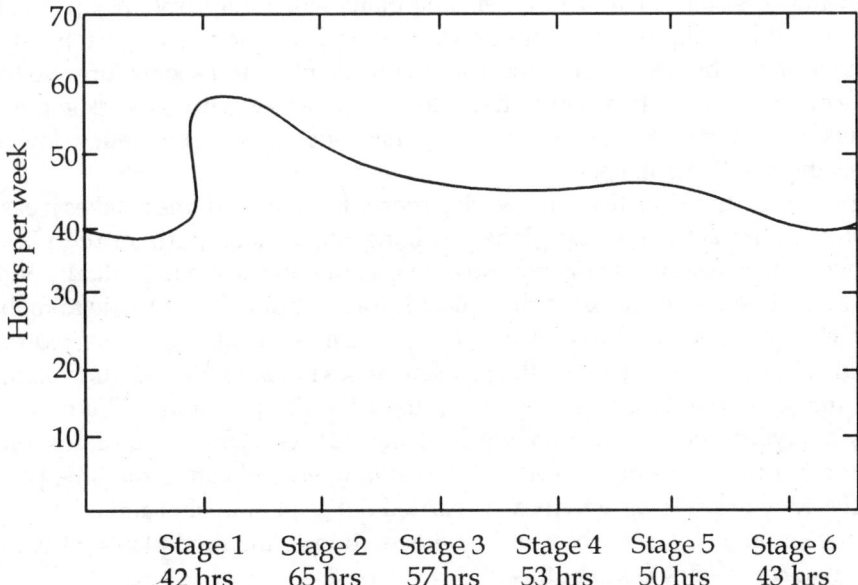

Fig. 7.3: Time Demands upon the Homemaker During the Different Stages of the Family Life Cycle. (Based on unpublished data, Bureau of Home Economics, U.S. Department of Agriculture).

An Aid to Time Plans for the Employed Homemaker

The modern times have seen another change in the use of the time by homemakers. Unlike her counterparts in the past, she has additional burden of gainful employment. When she is gainfully employed and goes out for work, then she has the dual responsibility of taking care of her career as well as home along with a possible third child rearing. And, her time demands are increased. If she is at home, then she can divide her timings for various activities and take out some time for rest and leisure whereas the gainfully employed homemaker is deprived of this pleasure of having enough rest and leisure.

The services used by the homemakers affect their time use. If there is home delivery of goods, laundry services etc. then she can save time. The use of labour saving devices also reduces the time of the homemaker thus giving her time for rest and leisure. Age, education and experience of the homemaker and her own attitudes towards work exert a great influence in the quality of work. If she is more experienced and skilled and has the right set of values and attitudes, she will carry out work more systematically and effectively. But, in lower income group Indian families, the time use pattern of the homemaker will be different. Here the homemaker cannot afford the paid help, rather she might herself be employed. Hence, she has to manage her time for both the household chores and the employed work.

CONTROL OF TIME PLANS

Carrying out time and activity plans to successful completion follows the making of the plan. Whether a plan is a mental record or a written one, it must be workable to be of any value. It must work when things are running along normally and with adjustment, it should work when emergencies arise. A good plan will always serve as a guide no matter what happens.

The homemaker's skill in adjusting her time plans when interruptions arise help to control many situations when illness occurs or outside work or shopping takes part of a day, it may be necessary to omit all but essential work from the day's plan. If it seems unwise to omit some tasks, a little more speed, shift in standards for the day, or an extra hours work in the evening may be the means of catching up. The best of managers cannot escape emergency days but they can learn to meet them with mental poise.

Establishing good time practices and work patterns for daily activities makes it easier to weave personal and leisure activities into plans. Bringing the various parts of the plan into a well controlled workable one, satisfactory control of one's time comes about gradually and easily .

It is just not enough if one makes a time plan. Efforts should also be made to follow it up. The plan, if written, should be kept at a place which is easily accessible and referred to . Conscious efforts should be made to check when the problem arises in following the time plans. Such plans should be made workable by finding the causes for the problems. The causes should be thoroughly analysed and the plan should be adjusted. As in the case of any other resource management, the two ways of controlling i.e. checking and adjusting the time plan need to be practised-checking to see what activities are carried out as per the plan and adjusting if there is a change in the situations. The successful implementation of the time plan will be possible only when these two control devices are utilised.

EVALUATING TIME PLANS

Evaluation enters into the making and carrying out plans, as well as reviewing the results . In planning, possible ways of reaching goals in time management such as meeting the time and work demands of daily living, satisfying the desires for leisure activities and mental growth and saving time from one task to use in doing another, one is constantly evaluating different lines of action in search for the best one .

As time and work plans are carried out there is constant evaluation of performance and checking of accomplishment to be certain that things are going about as planned. If time runs short, or if adjustment must be made in the day's work, the evaluation of many factors enter into the decisions that must be made. Conscious or unconscious evaluation of one's use of time and accomplishments, shapes and improves new plans. For the success of the time plan, evaluation is a flexible mental tool in management which is an important step to be carried out. Evaluation is a step done as the last step, in every managerial activity. While evaluating, one looks back at the plan made earlier and checks whether it has yielded the expected results and the goals are achieved. Besides checking for the attainment of goals, evaluation also serves as a feed back in the preparation of future plans.

TIPS FOR MANAGING TIME

Everybody has to manage their time to some extent whether it be at home or at work or both . The sequence in which you perform tasks on everyday basis has a profound effect on how much you get out of life. Most people have the capacity to manage their time better, and by doing so makes the working day more productive and leisure time more fulfilling. Time is one of the most valuable resource because it cannot be regained once passed. Therefore, it becomes more important to analyse time usage on a regular basis, it is possible to understand the most efficient ways to use time both at home and at workplace.

The following are some useful practical tips to manage time wisely.

- Set aside time each day to review and prioritise demands on your time.
- Take a small chunk of a difficult task, and deal with it straightway.
- Use planning and delegation to minimise time wasting at every level.
- Always delegate tasks which are not time effective for you to do.
- Estimate how long a task will take you, and see how accurate you were.
- Assess your abilities and identify your weakest areas . Learn to use your time efficiently and reduce the time you spend working in unproductive and labour intensive activities.
- Break down long term plans into weekly and daily action plans.
- Ensure that you have some quiet time everyday.
- Keep some energy for home and leisure activities after work. It is important to schedule relaxation time into your day, because your concentration levels and productivity start to decline as your tire.

Time, or rather lack of it, has always been a pressing problem. But, nowadays as people are becoming more work conscious, there has been more concern for time as a resource than ever before.

Managing the use of time has definite advantage, better appreciated by average person. Management of time involves the process of decreasing the time devoted for a particular work without sacrificing the quality of work. Time is the most perishable of all the resources. Since, it is a resource, individual or family must have a control over it. People face many problems, because of its limitations. Once they start realising the scarcity aspect of this resource they would attempt to improve the management of their time.

8

Energy Management ≪≪≪

THE energy resource shares many characteristics in common with the time resource. Everyone of us has a fixed amount of energy available for allocation and use each day. Like time, its replacement necessitates little of any exertion. Energy, is always combined with other resources in its allocation and use. Unlike time, quantitative aspects of this resource varies with individuals. Some have more than others. The supply of this resource may also vary somewhat. Some days one individual may have more energy than the other days. This variation in quantity also occurs not only from day to day but also within the day. There are certain times of the day when the same individual has more energy than the other times. One's energy resource is important to one's management. Knowledge of its availability and potential utilization are vital to everyone of us. This resource, combined with other resources need to be handled carefully within the management process.

Energy is classified under the human resource. As such its availability is governed not only by how one allocates and uses it but also by one's diet, hours of sleep, rest periods, and the attitudes towards work, time, etc.

Goals are important in energy management because they determine how much and what kind of energy should be mobilized and utilized. In making plans, two questions should be asked. How much energy should be spent in pursuit of a particular goal? How much is its attainment worth? If the cost in energy required to meet the goal appears to be too high, different goals may be set—goals within the supply of available energy, with enough energy left for enjoying and participating in other activities.

RELATION OF ENERGY TO THE STAGES OF THE FAMILY LIFE CYCLE

Although no studies have shown the actual demands on the homemaker's energy during different stages of the family life cycle, the demands discussed in the chapter on time management indicate the stages during which the most work must be done.

Energy demands are usually lowest during Stage I, the begining family stage. If the homemaker works outside the home, however, she must learn to divide her energy between the demands of her outside work and her home responsibilities.

The busiest years for the homemaker occur during Stage II, the period during which the family is expanding. As evidenced in the discussion of time management, young children require more care, and after the children start going to school, meeting all the family demands both inside and outside the home requires a great deal of energy.

When the children reach college age and begin to leave home, Stage III, the contracting stage, begins. As their energy loads lighten at home, many homemakers resume professional and outside interests. Helping married children, and welcoming new members to their families often brings additional responsibilities. During this stage, energy supply may diminish and physical disabilities may require the reshaping of energy spending patterns.

CLASSIFICATION OF EFFORTS USED IN HOMEMAKING ACTIVITIES

Performing each homemaking task requires several types and combinations of efforts. For instance, even performing any such routine tasks as dressing, sleeping, or dishwashing, which most homemakers do almost automatically, requires some mental effort. Although we are seldom conscious of it, most activities require visual effort because our eyes must direct the movements of our bodies. Muscular movement of the eyes and adjustment of vision to different distances and lighting conditions take place constantly. Such tasks as preparing meals, setting and clearing the table, dishwashing, and laundering require no small amount of manual effort— reaching, raising, lifting, holding, carrying, pulling, and pushing. Some of the more strenuous tasks, such as those connected with caring for the house, garden, and yard require torsal effort; that is, bending, leaning, rising, turning, stooping, sitting, and kneeling. And pedal effort— walking, moving, and standing—is an essential part of many homemaking and recreational activities.

Thus, while the various homemaking activities require different combination of efforts, most tasks require mental, visual, manual, and torsal effort of some kind, and many also require pedal effort.

Efforts used in homemaking Activities

- **Mental effort** involves the process of thinking, reasoning, planning, decision-making, directing, worrying, talking.
- **Visual effort** is related to seeing, it involves the transmission of visual stimulus along nerves to the visual cortex of the brain which enables us to see. It involves eye movements and pauses, looking, reaching, watching, adjustments to distance and lighting conditions.
- **Manual effort** involves physical work such as reaching, raising, lifting, holding, caring, stretching, pulling, pushing, meal preparation, laundry.
- **Torsal effort** involves bending, leaning, rising, turning, stooping, sitting, kneeling. (used in more strenuous tasks), gardening.
- **Pedal effort** involves the use of feet to operate or carry out a work such as walking, moving, standing etc.

Ways to Minimize efforts in Homemaking Tasks

As discussed above, efforts are required to accomplish various homemaking tasks which cause fatigue. However, fatigue can be reduced to a large extent by minimizing these efforts in most of the homemaking tasks. The different ways of minimizing the various efforts are suggested here to help the homemaker to become aware and utilize them effectively while performing the various homemaking tasks.

1. Mental Effort

Mental effort can be reduced by planning before hand the tasks to be accomplished, like thinking a rough outline of the task to be carried out and its possible outcome. Experience of handling household tasks and developing habits proves to be a great aid in minimizing mental effort. Besides, acquiring knowledge and getting familiar with the tasks also help a home-maker in reducing her mental effort, sharing experience and delegating work to others. Either family members or paid help can also be considered as good ways of minimizing mental effort.

2. Visual Effort

Visual effort can be reduced by keeping the things at the right height. Proper lighting arrangements—both general and local—should be maintained in the room. Using transparent containers and labeling the non- transparent ones, reducing the depth of the storage areas, storing things in their usual places and not frequently changing these, keeping the track of storage places for the seasonal items such as woollens etc. would also enable a home-maker to reduce her visual effort.

3. Manual Effort

Manual effort can be reduced by arranging storage so that much used articles are within easy reach without stretching. Heavy articles should be so placed that not much raising, lifting, holding or carrying is necessary.

The working surface should be large enough to allow for a good layout for the work process and the storage of equipment and supplies should be close at hand. Use of labour saving devices like washing machine, mixer grinder etc. also help in reducing manual effort.

4. Torsal Effort

Torsal effort is somewhat dependent upon manual effort. If the article in use are placed so that they can be reached and used with a minimum of bending, leaning, turning, stooping and rising then torsal effort is reduced. When the torso is used in any of these motions, the body is out of line and another part is extended to restore balance. Work should be planned to be performed on a platform rather than on the floor to avoid frequent getting up and sitting down which would help to minimize torsal effort.

5. Pedal Effort

Pedal effort can be reduced by avoiding static work and taking breaks in between work involving standing for long periods. Those equipments and appliances should be opted and used which would encourage more of sitting positions rather than standing. Comfortable footwear should be worn such as acupressure chappals and shoes which relaxes the foot and minimize the strain hence, the effort. Storing the items near the place of work would also involve less trips, thereby reducing pedal effort. Use of trays and trolleys for carrying a number of items, as in the case of serving food from the kitchen to the dining area, will also help a home-maker to reduce this effort.

Studies of Energy Costs of Household Tasks

To work out well-balanced energy spending patterns, homemakers need to know both the energy cost of various homemaking activities and those considered most fatiguing. The human energy required for the performance of any task is made up of several different parts. A certain amount of energy is needed for the maintenance of muscular tension and for the natural body processes such as respiration, circulation, secretion, digestion and excretion. This is known as resting metabolism. In addition there is the energy used in moving about and in the actual doing of the task.

In the studies that have been made, the energy expended in sitting, standing, walking, or in doing a task is measured by determining how much oxygen is consumed per minute. The results are given in calories used per hour for each pound of body weight and in percent above resting. In each experiment the resting metabolism of the person being tested, as well as the energy expenditure in performing the task, is determined. The difference between the total energy cost and the energy cost of resting gives the energy cost of the task itself.

In one of the first studies of the energy costs of a number of household tasks, the tasks were roughly divided into three classes:

- Light work, such as knitting, darning, sewing by hand and with a motor-driven machine.
- Moderate work, such as ironing towels, dressing infants, washing dishes, and sewing with a foot-driven machine.
- Strenuous work, such as washing towels and sweeping the floor.

The findings showed that light work required about 15 percent more energy than resting in a chair, that moderate work required about 24 additional calories per hour and the strenuous work required about 50 additional calories per hour.

Household Tasks Classified by Energy Costs

The household tasks mentioned in the various studies have been classified according to energy costs as light, moderate, and heavy, and are given in Table 8.1. This covers a wide range of activities and includes many of the tasks most frequently done in the home. Moderate or heavy work requires both walking and standing as well as different forms of manual and torsal effort. Checking the energy costs as given in the list in the following paragraphs against the different forms of the effort used in their performance will help the home-maker to select light, moderate, or heavy energy demanding tasks to make a comfortable daily and weekly work load.

What Activities Are Most Tiring?

Measuring energy cost of doing household tasks, though useful in planning both work and household design, gives little indication of the fatigue the worker experiences, because different kinds of work have different fatiguing effects. Each of us responds to work differently, and fatigue patterns vary in many ways. Some light tasks that involve little expenditure of energy may be very tiring because of mental approach, postural strain, muscle tension, or the concentration and skill they require, whereas some heavy work that requires more energy may be far less fatiguing than some light work.

TABLE 8.1: ENERGY DEMANDING TASKS

Light:1.4-2 *Calories per minute*	Moderate:2-3.5 *Calories per minute*	Heavy:3.5-4.5 *Calories per Minute*
Hemming	Using carpet sweeper	Scrubbing floor
Knitting	Using vacuum sweeper	Mopping floor
Crocheting	Polishing furniture	Waxing floor
Darning	Kneading dough	Taking out and hanging laundry
Hand sewing	Wringing clothes with	Washing kitchen floor
Machine sewing	electric wringer	Bed making
Preparing meals	Hanging clothes from buckets	Lifting heavy buckets of wet
Washing dishes	cleaning utility table	clothes
Dusting furniture and Floors	Ironing	Lifting young children
Sweeping kitchen floor		

The tasks that homemakers considered most tiring have been reported in a number of studies. Many homemakers indicated cleaning and caring for the house and washing and ironing tasks were most tiring and the tasks they disliked most.

Because of lack of efficient equipment and dislike towards the work were listed as the main causes of fatigue. It is apparent that providing work saving equipment and finding ways of making work more enjoyable and surroundings more attractive can reduce fatigue costs. Some methods of doing these will be discussed later in this chapter.

In this fast world, where everyone wants to utilise even a second of his or her time and energy to convert them into money for his or her family's well-being, the use of both the resources time and energy are of very great importance. Clear understanding on how energy can be managed is of importance, since time and energy are related resources, especially for the homemaker who has to go out for work, besides working in the house. Therefore when time is utilized, we automatically use energy resource also. Where more time is spent on a task, it also means a higher expenditure of energy resource. Similarly when the time spent on a job is reduced, it would also automatically reduce the expenditure on energy. However, there are some independent factors that are exclusively responsible in the management of energy. For proper utilization of energy, a few factors that are directly involved in energy management are discussed under body mechanics.

Body Mechanics

Among the important aspects of energy management, and one of growing concern, is that of body mechanics. It is *a term gradually replacing posture because it includes the body in motion as well as in standing and sitting positions. The word 'mechanics' suggests a relationship to the functioning of the body.*

As was aptly stated by Esther Crew Bratton: *Your own body constitutes your most important item of household equipment. It is well worth the effort it takes to acquire understanding of how the body functions in work and to develop skill in using the body effectively. Nor is it a tool that will change in the foreseeable future!*

Body mechanics includes feelings of comfort and discomfort associated with the use of muscles and skeleton. Among the major principles in their use are:

- Keeping the body parts in alignment.
- Using muscles effectively.
- Rhythm in movements.
- Considering the centre of gravity, both of the body and of articles handled.
- Taking advantage of momentum.

These five aspects are interwoven in actual uses of the body and its parts.

Keeping Body Parts Aligned: Keeping body parts in alignment results in stability when the various body weights are correctly positioned, each centred over the base of support. (Figure 8.1)

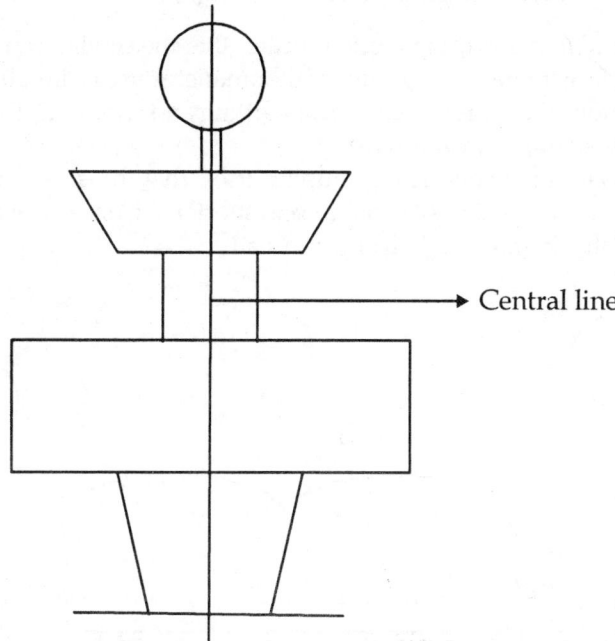

Fig. 8.1: Alignment of Body Parts

When any part of the body gets out of line, muscular effort is required to maintain body balance in addition to whatever work the body is doing, thus extra strain results. Therefore, correct or the right posture should be maintained by the person. For the correct posture whether sitting, standing or using a tool, muscles are so constructed specially to do certain things or does the job that no strain is felt by the person, while in incorrect posture, muscles which are not constructed for that particular job, do the job, and therefore strain and tension is felt by the worker. When the strain is felt, extra energy is spent. Hence, for correct energy management, it is necessary to maintain the body parts aligned properly.

Using Muscles Effectively: Using muscles effectively includes employing,

- The strongest muscle feasible,
- Setting the muscles that are to do the work before contacting the load,
- Contracting muscles slowly, and
- Using muscles rhythmically.

For example, leg muscles, being stronger than back muscles, should be used for lifting loads. This is accomplished by standing or kneeling close to the load and lifting with a slow, steady pull.

(a)   (b)

Fig. 8.2: Postures for Lifting a Heavy Load (a) Incorrect (b) Correct.

Study of muscles used in performing work indicates that the smaller muscles become fatigued more quickly than the larger ones, though use of the smaller muscles involves the expenditure of less energy. Hence, in household processes, such as cutting and chopping, the arm muscles should be used instead of those of the hand and wrist.

One more point to be noted concerning muscles itself that, muscles exerts its greatest force when extended, and force diminishes as the muscle shortens. Hence, when lifting a heavy sofa, the legs should be slightly bent before grasping the load.

(a)   (b)

Fig. 8.3: Stretched muscles exert greater force: Two Persons Holding Sofa From Each Side, Using Muscles (a) Correct.(b) Incorrect.

Rhythm

Rhythm in muscular performance may be defined as the repetition of movements at the same tempo. In the rhythmical activity, a large part of the first excitement still serves for the second, and the second for the third and so on. Inhibitions fall away, and the mere after effect of each stimulus secures a great saving for the new impulse. Thus it saves energy in regular rhythmic movements.

Bratton explains why rhythmic work is less tiring than non-rhythmic, as based on the existence of double sets of muscles for accomplishing work. When they work rhythmically, one set rests while the other set works. In non-rhythmic work both sets may operate at the same time thereby making the work more tiring.

In the household many activities can be done rhythmically. For example, in dishwashing, when all the plates of one size are washed in sequence, the hands and arms develop certain rhythmic movements in handling them. Sweeping, running the vacuum cleaner and ironing clothes, all provide recognized possibilities for a regular flow of motions, resulting in rhythmic movements. In this manner, rhythmic motions help a great deal in managing energy effectively.

Centre of Gravity

Considering the centre of gravity is of importance in lifting, supporting, or carrying a load and also in reaching to get an object. It is always desirable to keep the load close to the body so that centre of gravity is taken care of.

Fig.8.4: Lady Lifting the Child from the Ground (a) Incorrect Posture (b) Correct Posture
(Substitute Leg Muscles to Back Muscles)

Note in the above figure, how much closer the baby is to the women's body is good as compared to the poor method of lifting. The custom of carrying the baby on the mother's back, as the Japanese do, is an example of keeping the load close to the body, thereby walking along the centre of gravity.

Secondly, as far as possible, keeping the centre of the weight of the object through the centre of the body, and avoiding twisting the body helps in better energy management. Applying force to the centre of gravity of a load to be moved is the economical use of energy.

(a)   (b)

Fig. 8.5: Lady Pushing an Almirah (a) Incorrect Posture (b) Correct Posture
(Use Whole Body at Centre of Weight to be Moved)

Taking Advantage of Momentum

Taking advantage of momentum means the avoidance of stops and starts, and of change of speeds. Similar to rhythmic movements, free flowing motions are the least fatiguing of movements because they are continued so that one motion flows smoothly into the next, rather stopping abruptly. For instance, in polishing or dusting a large surface, the end of each movement may be rounded to make the return stroke a continuation of the forward stroke so that momentum of the work is maintained.

POSTURE IN HOUSEWORK

As discussed earlier, we can conclude that to avoid strain and to develop a good body carriage while working, some attention should be given to posture habits in standing, sitting, stooping and bending while at work. Good posture in doing any task may be defined as the position which requires the expenditure of the smallest amount of energy. A good standing posture is one in which the head, neck, chest and abdomen are balanced vertically one upon the other, so that the weight is carried mainly by the bony framework and a minimum of effort and strain is placed upon the muscles and ligaments. When the body is well balanced in the standing position, the head will be directly over the feet, and the center of gravity will pass through the middle of the face, nose, shoulder, hip, the outside of the knee, and the outside of the ankle.

A good sitting posture for work is well- balanced and poised position. The weight is carried by the bony support of the skeleton thus relieving the muscles and nerves of all strain. The poise is such that minimum adjustment is necessary for such action as the work may demand. The line of gravity falls through the middle of the shoulders, hips and seat bones. The body is straight from hips to neck, and there is no flex or bend at the waistline.

Poor standing and sitting postures may result in permanent changes in the spine in positions of the joints, ligaments and muscles and in the location of the organs of the body. Such changes produce strain and tensions which increase the fatigue costs of homemaking tasks.

Using the most comfortable body position while working eases the body and relieves strain. Alternating standing and sitting is more restful than either one continued for a long period.

Doing a task in the efficient way means saving both time and energy. Hazelton and Russell describe the efficient way as *the one that uses only those muscles or parts of the body that are necessary to the operation and uses those muscles or parts of the body which are best able to do the work.*

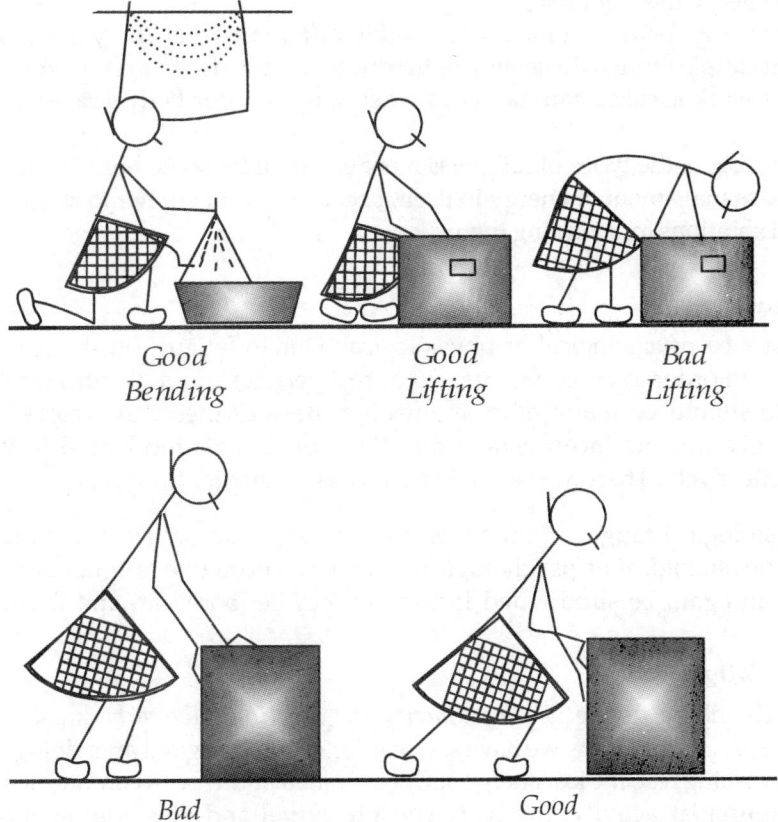

Good
Bending

Good
Lifting

Bad
Lifting

Bad

Good

Fig. 8.6: Pushing and Pulling, Good and Poor Postures for Some Household Tasks

For instance, in bending to do certain tasks, it is easier to put one foot slightly forward and bend through the knee and ankle joint. In lifting something heavy from the floor, such as a body or a bag of groceries, it is better to bend the knees and thigh joints and use the stronger leg muscles for the actual lifting. Pushing or pulling a large piece of furniture can be done with least strain by dropping into a partial crouch, knees limber, hips low, and pulling or pushing in that position. Carrying packages or other articles is easier when the load can rest against the hip. Balancing is done by bending toward the other side which shifts the weight to the large bones and leg muscles. That is why we find most of the labourers carrying their load at their backs and resting them on their hips.

FATIGUE; TYPES, CAUSES AND REMEDIES

Keeping in mind the above factors while managing energy, still one may feel fatigued. Why does this happen. For better energy management one should know how fatigue happen and their causes so that one can make efforts to reduce them.

Along with the revision to have the complete idea about fatigue and some concepts of the use of energy, there are come changing ideas about fatigue. Recent research in physiology, psychology, industrial management and home economics has greatly increased information on fatigue. It is no longer considered a unified thing but is classified into types. An understanding of different forms of fatigue and their influence on mental and physical efficiency is essential in handling problems connected with energy management.

Fatigue is not easily defined. It manifests itself in different forms and yet they are the closely related forms. In simple words, the feeling of tiredness and the desire to stop working after doing some amount of work is called fatigue. Thus when fatigued, our body's capacity to do work is reduced.

An understanding of the types of fatigue is more essential for some homemakers in a practical sense than is the management of energy in itself, because it will help her to analyze the causes of fatigue and find solutions for avoiding them.

Types of Fatigue

Fatigue can either be psychological or physiological. Due to fatigue, our body's capacity to do work is reduced. In order to avoid fatigue, either rest period should be introduced between the work or the job should be made more interesting. Besides these, the worker can either be appreciated or given some incentive, so that the worker gets motivated to do work more efficiently and effectively. There are basically two types of fatigue. They are:

- The physiological fatigue: Sometimes also called tissue physical impairment.
- the non-physiological or psychological fatigue or subjective fatigue: This psychological fatigue can again be subdivided into two types i.e. boredom and frustration.

Physiological Fatigue

Haqqard and Greenberg define it *as incapacity for exertion induced by previous exertion. This capacity is restored only by rest*. As we have seen earlier, we feel fatigued after doing work and want to take rest. After taking rest, the lost energy can be regained and we can continue to work further.

During the muscular activity, the body consumes fuel and gives out energy. The energy producing material in the muscle is mainly glycogen, which is formed by muscle tissue from sugar products brought to it by the blood. In muscular work, glycogen unites with oxygen in the blood stream, releases energy and forms lactic acid and carbon-di-Oxide (CO_2). Both these waste products interfere with continued muscular activity of the body.

This state of the body results in a feeling of tiredness which is termed as physiological fatigue. The process of accumulation of lactic acid and CO_2 can be represented as:

$$Glucose + O_2 = Energy + Lactic\ Acid + CO_2$$

Recovery or the removal of lactic acid and CO_2 in the muscles is necessary after any and every kind of work. The blood stream picks up CO_2 and carries it to the lungs where it is exhaled. At the same time, the blood brings Oxygen to the muscles, and lactic acid is oxidized and reconverted to glycogen. Thus oxygen helps to prevent fatigue by aiding removal of lactic acid in the muscles. However, this can be done only if the body is at rest, and does not release any more lactic acid. If the body continues to work, the oxygen supplied will be used up by glycogen for the release of energy, making the oxygen unavailable for the removal of lactic acid from the muscles. That is why a person gets totally exhausted while doing continuos work.

The amount of work done during a day may be shown by a work or production curve. Unlike industry in which there is a morning and an afternoon work curve in homemaking, the daily work is varied and more or less continuous and may be expressed in a single curve as shown in figure 8.7

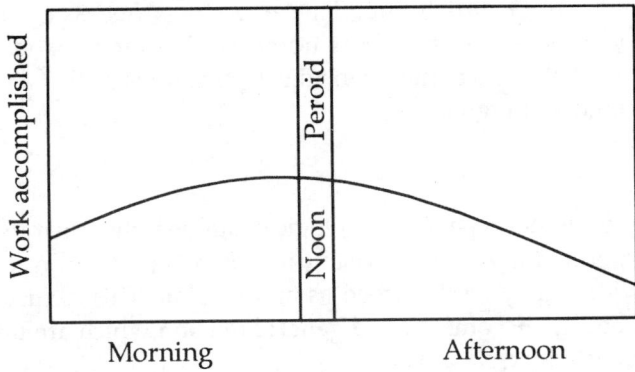

Fig. 8.7: Daily Work Curve of the Home Maker

According to the work curve, the homemaker's working capacity rises to its maximum height during the early hours of the morning and gradually declines as fatigue increases. Certain factors like illness, a poor night's sleep, nervous tension, an unusually heavy day's work, or over fatigue may cause variations in the shape of the curve from time to time. Each homemaker, however, has her own individual work pattern that remains fairly constant from day to day. A homemaker can learn to plan for the best use of her energy as well as the energy of those who help her with the day's work if she knows her own daily habits, feelings and how she tires. She must plan her work in such a way so that she has some rest period in between. This is very important because after taking rest, lost energy is regained and it is made available to do further work. Rest alone can reduce the physiological fatigue felt by a worker.

Non-Physiological or Psychological Fatigue

Sometimes fatigue may occur even when there is enough energy to do work. This will also result in reduced output. This is called as psychological fatigue. Just as physiological fatigue results in a reduced output of work and aversion to work, similarly, psychological fatigue too results in an aversion to work further and reduced output. It is very difficult to define or measure, as it consists in rather vague subjective states. It is not localized in the body. Differences in fatigue feelings due to **boredom** and due to **frustration** have yet to be classified.

Boredom Fatigue

In boredom fatigue there is **discontent, yawning, restlessness** and a **desire to** *quit* but the actual capacity to do work is not changed. Wyatt and Longdon found that those workers who were most frequently bored made the greatest number of complaints concerning working conditions. It is true that boredom increases due to certain objectionable features connected with work, such as noise, atmospheric conditions and troublesome materials and tools. Another symptom of boredom consists in the estimates of time at work. To the bored person, time seem to pass slowly, and this itself results in boredom fatigue.

A study by Ryan, with workers engaged in sewing operation showed that susceptibility to monotony on the job is related to dislike of regular habits outside of working hours. Those who adjusted well to the work tended to prefer a regular routine of household work and outside activities. The monotony susceptible worker tended to be more poorly adjusted to her relationships with her family and home. Therefore, it is necessary that the housewife understands this aspect, because her job at home is also more or less monotonous. An understanding of this will help a homemaker to adjust her routine so that a variety is created and she will not undergo boredom fatigue.

Frustration Fatigue

Besides boredom fatigue another type of psychological fatigue is the frustration fatigue. This may be regarded as an experimental pattern arising in a conflict situation in which the general alignment of the individual may be described as an **aversion**. This particular pattern involves **feelings of limpness, bodily discomfort**, and **general tension** which are undesirable as well as result in inadequacy for activity.

These feelings of discomfort and aversion to work should not be considered as symptoms of fatigue but they are the fatigue itself.

When plans fail to work out and goals cannot be reached or when conflict situations arise which call for the weighing of alternatives in the making of decisions, and the seeking of new goals, a person may experience feelings of frustration and increased tensions. Fatigue which results from such conflicts is a part of the total picture of frustration. Therefore, this kind of a fatigue is termed as *frustration fatigue*.

Knowles found some evidence of frustration fatigue in her discussions with homemakers regarding their attitudes toward certain tasks. Some of the reasons given for disliking tasks were:

- Uncertainty and confusion in performance
- Conflicting standards within the family group and inability to satisfy all members.
- Unfavourable working conditions.
- Practices and standards of work conditioned by tradition and in conflict with new developments and methods, and,
- Time required by the task could be used for other more interesting activities.

Thus, we find that there are a number of reasons for frustration fatigue. However, they can be broadly put under three categories. In the first instance, frustration fatigue may arise from the job itself. For example, the inexperienced person whose first pie or first slip cover is something short of success and which might take longer to be completed, may feel utterly fatigued.

The long period of work may be the result of interruptions and clutter and so the worker will complain of the feeling of fatigue.

Secondly, fatigue may arise from competition between what one is doing and what one would like to do. The homemaker working at home may feel fatigued when she know that kitty party is in full swing next door which she wants to attend but due to loads of work at home she cannot, so she will complain of fatigue and try to avoid doing the work at home.

Finally, fatigue may arise from something entirely unrelated to whatever one is doing. For example, the homemaker working at home may feel fatigued as she know her child has not arrived back from the school or a serious illness in the family may make her simple household jobs very fatiguing.

HOW TO AVOID FATIGUE

It is now clear that there are two types of fatigue—physiological and psychological, which a home-maker may face in her work. However, she can take certain steps to remove or avoid these so that she can continue to perform her work without feeling fatigued.

Rest Periods

The first method of removing fatigue is introducing rest periods in between the work. As is said earlier, regardless of its type, rest periods are of value both for increasing output and for alleviating fatigue. The kind of rest period should be related to the kind of fatigue, to be most effective. The effectiveness of a rest period is related to the completeness of relaxation. Relaxation comes at different rates to different people and may never be complete for same people.

To alleviate physiological fatigue, there must be cessation of physical activity to allow time for the fatigue products to be removed from the body and in the cases of greater fatigue, time for the store of glycogen to be replenished. Only in sleep the body can entirely eliminate physiological fatigue.

There is frequent confusion between local muscle discomfort and fatigue. Change of task will lesson this discomfort. If a housewife is required to perform an exceedingly painstaking task for only a part of the day and is shifted to work which requires much less concentration and the use of different muscles, the output is found to improve. The homemaker is responsible for a variety of tasks and may easily alternate types of work. Since energy expenditure is still occurring at a normal rate, this type of rest period cannot be expected to eliminate much physical fatigue but is satisfactory for both boredom and frustration fatigue.

Putting Interest into the Job

In reducing boredom fatigue, it is wise to remember that small changes in the task may be helpful, or that interest outside the job may be introduced, such as listening to music. Mayo suggests that a job may be made more interesting by the simple device of setting intermediate goals which are easily attainable, thus avoiding the seemingly endlessness of the job ahead. For instance, the woman who tries to complete the stacking and then sets another goal, feels a glow of accomplishment and a stimulus to speed more quickly than the person who hopes to complete the dishwashing in thirty minutes, but does not know until the end whether or not she has succeeded. Another suggestion for making tasks more interesting is to work in groups. While one of the characteristics of the homemaker's work is her isolation, there are possibilities for making the home tasks made more interesting if some of the jobs can be organized as a group rather than as individual ones. Laundering, can be more pleasant if one person carries the clothes to the line while another person prepares a fresh tubful for hanging.

An analytical Approach to a Job

This comes through changed procedure and actually reduce the physiological fatigue and at the same time create interest in the work itself. For instance, a young girl will take more interest if she has to prepare the entire dinner including the entire dessert course than if she is asked to mash the potatoes, slice the bread, pour the water and do numerous other routine jobs. Besides these, there are more ways to remove the feeling of fatigue and better energy management.

Having Proper Work Place and Proper Equipment

Such as good environmental conditions, proper equipment like the table according to the homemaker's height, adjustable shelves in the kitchen, etc. can also aid in reducing fatigue.

Appreciation of the Work

When the home maker is appreciated by the family members for her work, it motivates her to do the work more effectively, without any complaints of feeling fatigued. Appreciation also acts as an incentive for the worker.

Giving Incentive to the Worker

The homemaker can give incentive to her children to do work, like promise them that she will take them out if they help her in the work. Similarly, the worker herself can find incentives for herself, like an ice-cream, an outing to meet friends, etc. while deciding to take up and finish the work within the specified period.

REMEDIES FOR FRUSTRATION FATIGUE

For frustration fatigue the only genuine remedy is to seek out the underlying cause of frustration and if possible, remove it. However, the homemaker worrying over the fact that her husband is in a precarious occupation may find that once she accepts this fact, the frustration will be lessened. As a matter of fact, the mere recognition of the source of the trouble will disassociate the weariness and the work, and an individual may then actually find his work a way of *Losing himself*. One the of most frequently disturbing factors inside or outside the job is unsuccessful personal relations with others. In this regard, one must remember that it is possible to increase one's proficiency in living and working with others. The rest period may be helpful at least temporarily in lessening frustration fatigue—as, for example, turning from the family account book which does not balance, to take a walk to the store for groceries and then come back to have a fresh look at the accounts. Complete physical rest may also be of great value in relieving tension.

Role of Motivation

Studies that have been made in industry show that level of motivation has a relation to all forms of fatigue. When an individual dislikes a task, or when motivation is at a low level, fatigue becomes apparent very soon but when motivation is high, fatigue may not be apparent until considerable exhaustion is shown. According to Maier the motivating condition in the work situation help to determine the amount of energy one has to spend for a task. High motivation appears to make more energy available for the task to be done, while low motivation releases less energy. We may conclude that motivation thus plays an important role in fatigue costs of homemaking activities.

All work is made easier and more interesting if goals are involved. Planning immediate goals which are easily reached makes work less monotonous and steps up motivation. It helps to break a big job down into several smaller parts. The feeling of satisfaction in completing each small part helps one to reach the final goal. Interruptions often interfere with the completion of tasks, disorganizing the day's work and causing much fatigue. Unfinished tasks leave a feeling of frustration or inadequacy in the mind of the worker. It is easy to forget the finished tasks, but hard to forget those that are left unfinished. Allowing free time is the daily work plan is one way of meeting this situation. It may be that the homemaker who straightens the living room before retiring does this because of the satisfaction it gives her in finishing the day's work and in being ready to meet the next day.

Some kinds of fatigue come and go. Fatigue that vanishes with a little excitement, an invitation to a movie or a picnic, a change in routine activities, or a new interest may be the result of boredom or minor tensions.

TECHNIQUES OF WORK SIMPLIFICATION

Anyone who is trying to lower time and energy expenditures soon learns the value of improving methods of work, because the time and energy required to do any task depend largely on the hand and body motions used. Improvement in the performance of a task usually means that the work is made easier because the new method is a more convenient one, permitting smooth, natural and rhythmical motions.

There are two time and work reducing ideas — work simplification and motion mindedness — may be used by everyone. Work simplification is the conscious seeking of the simplest, easiest and quickest method of doing work. Motion -mindedness is an awareness of the motions involved in doing a task and an interest taken for the possible ways of reducing them.

Attention was first focussed on work simplification through research carried on in the industrial field. Motion and time studies showed that improvements in methods of work not only eliminated useless motions but also saved the worker's time and energy. Work simplification research consists of making motion and time studies of the work as it is being done, analyzing the work methods, developing the easiest and most effective way to do the task, and putting the new method into use.

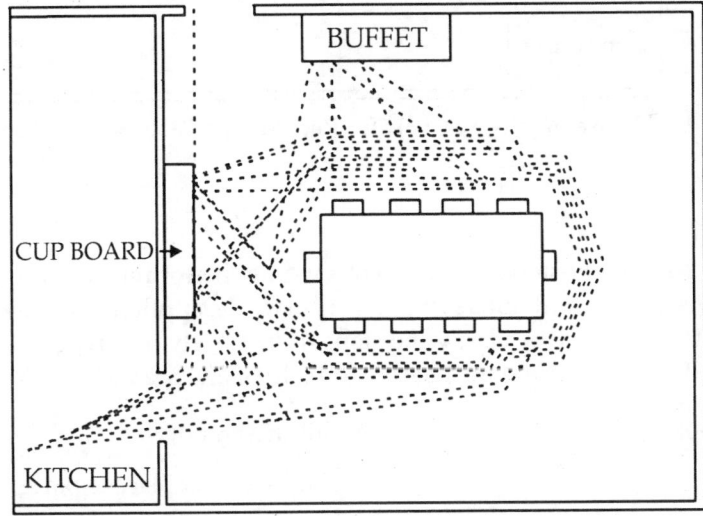

Fig. 8.8: Original Pathway Chart for Recording a Table Setting Activity

Some of the techniques used for motion and time studies are:
- Pathway chart
- Process chart
- Operation chart
- Micro-motion film analysis.

Pathway Chart

The **Pathway Chart** is a simple device for making a motion and time study in the home. A floor plan drawn to scale and fastened to a drawing board or wallboard, pins and thread are all that are needed to make such a study.

Thumb tacks or pins are put in on the floor plan where the worker turns and her pathway is measured from thread wound around the pins as she works. Two charts are used - one for the original and the other for the revised pathway (Figures 8.8 and 8.9). If measured lengths of string are used, the task of figuring distances traversed is much simplified. Because of tangling when removing any amount of string left over can be more easily measured than the total amount used. If not enough was measured off at the start, additional measured amounts can be used.

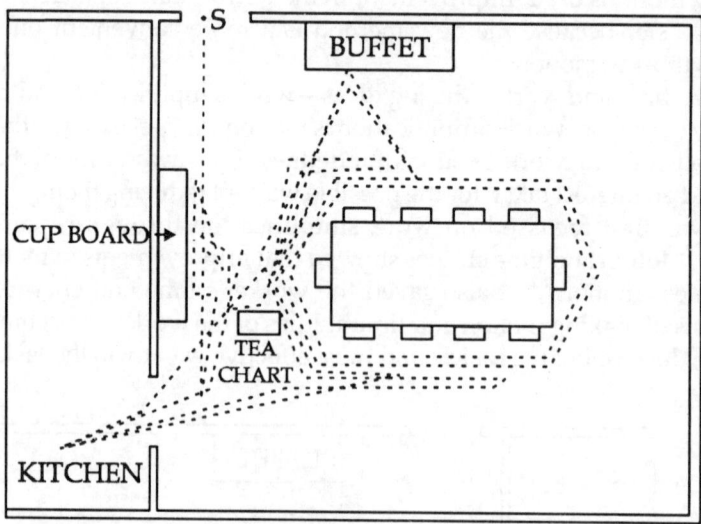

Fig. 8.9: Revised Pathway Chart for Recording a Table Setting Activity

Process Chart

The **Process chart** is a step-by-step description of a worker performing a given task in its entirety. It is an over-all investigation and differs from an ordinary description of a worker's activities only in that a few symbols are used to clarify the steps immediately into types. Gross, Crandall and Knoll have adapted some of these symbols from industry in the preparation of a process chart.

Symbol	Name of Symbol
◯	Movement from place to place
⬭	Operation
▢	Inspection for quantity or quality
▽	Delay
◎	Movement and operation done simultaneously

The small circle indicates that the worker is going somewhere; the large circle indicates that she is standing still but working with her hands; the square indicates that she is checking what she has done; the triangle indicates that nothing is happening; the composite symbol indicates that she is accomplishing something with her hands while walking. Thus, in charting the setting of a table, each time the worker walks is indicated with the small circle. When her hands alone are working the large circle is used. When she stands still but for example looks over her job to see if it is completed, the square is used. It can be distinguished from delay by the movement or focus of the eyes. The advantage of using the symbols along with the description is that one may quickly count up the number of each type of steps. A composite symbol counts as two activities when summarizing.

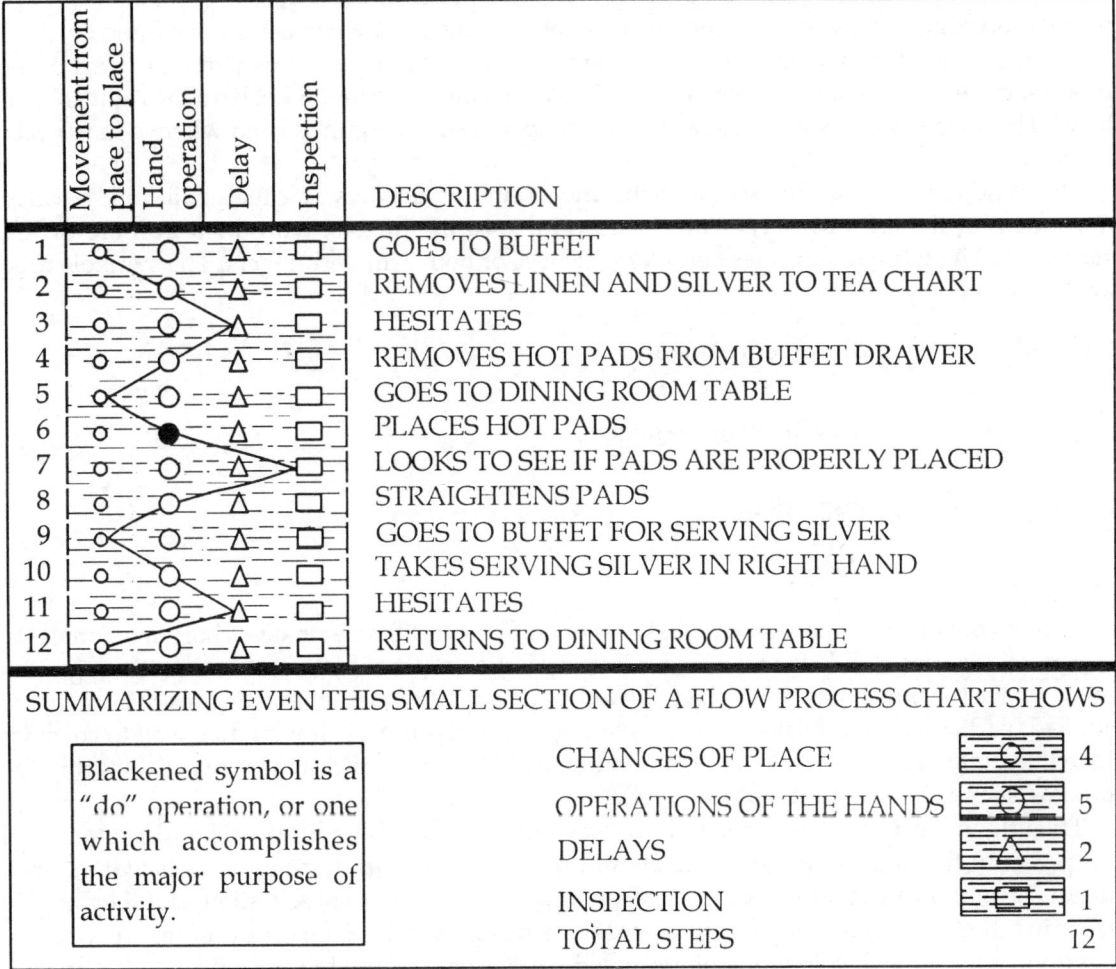

	Movement from place to place	Hand operation	Delay	Inspection	DESCRIPTION
1					GOES TO BUFFET
2					REMOVES LINEN AND SILVER TO TEA CHART
3					HESITATES
4					REMOVES HOT PADS FROM BUFFET DRAWER
5					GOES TO DINING ROOM TABLE
6					PLACES HOT PADS
7					LOOKS TO SEE IF PADS ARE PROPERLY PLACED
8					STRAIGHTENS PADS
9					GOES TO BUFFET FOR SERVING SILVER
10					TAKES SERVING SILVER IN RIGHT HAND
11					HESITATES
12					RETURNS TO DINING ROOM TABLE

SUMMARIZING EVEN THIS SMALL SECTION OF A FLOW PROCESS CHART SHOWS

Blackened symbol is a "do" operation, or one which accomplishes the major purpose of activity.

CHANGES OF PLACE	4
OPERATIONS OF THE HANDS	5
DELAYS	2
INSPECTION	1
TOTAL STEPS	12

Fig. 8.10: Process Chart for Recording a Table Setting Activity

It requires at least two persons to make a process chart, one to do the task, and at least one other to observe and record. The time is relatively unimportant, as the focus is upon the flow of work. The chart helps to visualize the sequence of an activity. For home tasks, the worker is followed throughout and the chart may be called a process chart- man analysis. In industrial process charts, sometimes a product, not a given worker is followed, but only the person or the thing can be charted at one time not both. It is customary, in this method of research to perform and chart the same task in an original and then in a revised way. The count of symbols after the original way often indicates at a glance where improvement may be made. An example of a portion of a process chart is given in figure 8.10. Looking toward improvement, one would immediately question four changes of place and two delays in only twelve steps, and attempt to eliminate some or all of them. Some delays are avoidable and some are not.

Operation Chart

The **Operation Chart** is similar to a process chart, except that it picks up one particular step in a whole process and breaks it down into the work of each hand, shown in parallel columns.

The Operation Chart is used in making a more detailed study of some particular part of the process. In this chart, the movements are broken down into the activities of both the right and left hand. The finer analysis show where necessary motions are being made and where delays occur in work.

The same symbols may be used, with the understanding that this time the small circle means a movement of the arm, the large circle a movement of the fingers, with the arm more or less stationary. The triangle indicates complete idleness of both arm and fingers. The symbols used are:

○	Movement of the arm
◯	Movement of the fingers
☐	Inspection
▽	Delay

It takes considerably more skill to make an operation chart than a process chart- man analysis. It is practically impossible to make an accurate operation chart of two hands except through film analysis. Each hand can, however be followed by a separate observer and actions of both hands studied, particularly for delays. An example is given in figure 8.11. It would be well to consider here if the idle hand (especially the left hand in step 7) would perform a more active part in the process, such as turning the potato as pared.

Thus the main uses of process and operational charts are for educational and promotional purposes. In this they serve an extremely useful end. They help in training investigators, they make one motion and time conscious, they sharpen one's power of observation, and they help in learning the principles of effective work. In other words, charting creates an orderly environment for the development of improved work methods, but in the final analysis, much of the creation is the product of the analyst's ingenuity.

Micro-motion Films Analysis

Micro-motion film analysis is primarily a research technique and applies best to tasks that can be easily filmed. Motion pictures of tasks done under normal conditions make a permanent record that can be analyzed and charted to show the work of the hands or other parts of the body used in the operation. By means of a timing device, the time of each movement of the worker can be accurately recorded.

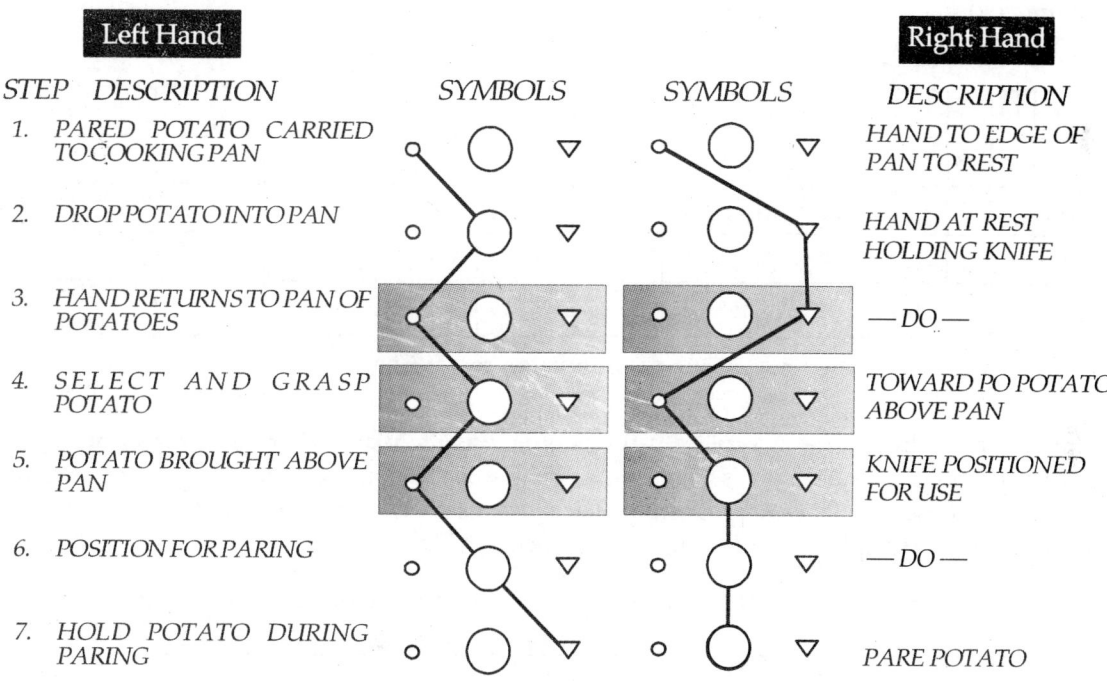

| | Left Hand | | | | Right Hand | |

STEP	DESCRIPTION	SYMBOLS		SYMBOLS		DESCRIPTION
1.	PARED POTATO CARRIED TO COOKING PAN					HAND TO EDGE OF PAN TO REST
2.	DROP POTATO INTO PAN					HAND AT REST HOLDING KNIFE
3.	HAND RETURNS TO PAN OF POTATOES					— DO —
4.	SELECT AND GRASP POTATO					TOWARD PO POTATO ABOVE PAN
5.	POTATO BROUGHT ABOVE PAN					KNIFE POSITIONED FOR USE
6.	POSITION FOR PARING					— DO —
7.	HOLD POTATO DURING PARING					PARE POTATO

Fig. 8.11: *Operation used for recording Potato paring Activity*

The cycle-graph, a photographic device, is also used to study types of motions used in performing tasks. When this is attached to some portion of the body, such as the hand when ironing is being done, it registers the pathway of light projected by a small electric bulb. The resulting record shows whether the movements are smooth and rhythmic or non-rhythmic. This is an effective way to learn how motions may be reduced and how methods of work may be improved in doing a task.

Path Process Chart

A new technique named Path Process Chart was developed by Magu P., Seetharaman P., Khanna K.[1] for conducting a study on (Urban Kitchens and Meal Preparation Activities- An assessment of their association (2000). It consists of a chart on which the activity, as being performed by the worker, is recorded. Later, the data can be analyzed using especially designed analysis sheets, The format of the new technique is shown in Figure. 8.12. It consists of at least 20 columns. These represent the following:

---

[1]Magu, P. Seetharaman, P. and Khanna, K. (2000). *Urban Kitchens and Meal Preparation Activities – An Assessment of their Association*. Unpublished doctoral thesis submitted to the University of Delhi, Delhi.

Columns 1 – 13   Workplace address
Columns 14 – 17   Process
Columns 18   Time Duration
Columns 19   Equipment
Columns 20   Remarks

⌐—WORKPLACE ADDRESS—————————→											⌐ PROCESS ⌐						
Do	*S	S	S*	*R	R	R*	RF	EA	1	2	3	AO	O	T	D	S	TIME	EQPMT.	REMARKS
Do	*S	S	S*	*R	R	R*	RF	EA	1	2	3		O	T	D	S			
Do	*S	S	S*	*R	R	R*	RF	EA	1	2	3		O	T	D	S			
Do	*S	S	S*	*R	R	R*	RF	EA	1	2	3		O	T	D	S			
Do	*S	S	S*	*R	R	R*	RF	EA	1	2	3		O	T	D	S			

Fig. 8.12: Format of Path Process Chart

1. Workplace Address: The workplace can be defined as the place where the worker actually performs the tasks. The address is given on the basis of the major equipment present at that workplace or present in its close proximity. The symbols used in the chart and the workplace denoted by them are explained in Table 8.2 and depicted in Fig. 8.13.

TABLE 8.2: SYMBOLS USED IN THE PATH PROCESS CHART FOR DENOTING DEFFERENT WORKPLACES

Symbol	Workplace Denoted
Do	Door
S	Sink
*S	Workplace towards the left of the sink
S*	Workplace towards the right of sink
R	Range and whatever area is left in front and behind
*R	Workplace towards the left of the range
R*	Workplace towards the right of range
RF	Refrigerator
EA	Eating area i.e. any table, could also be a dining table, used or meant to be used for the purpose of having meals.
1, 2, 3...	These are the workplaces on the long, continuous counter of the kitchen where no major equipment has been placed.

O-Operation: It is a sequence of activities taking place at a workplace during which the characteristics of the material are changed. e.g. cutting of vegetables, cooking of rice, washing of utensils etc.

T- Transport: This represents the physical movement of the worker from one workplace to the other.

D-Delay: This represents an interruption in the carrying out of work. It could be inherent in the work itself (ID i.e. inherent delay) e.g. waiting for the '*masala*' (onion and garlic paste) to become brown before adding water for gravy. It could also be intentional delay (Int. D) e.g. intentionally *leaving all work to talk to a family member. The delay can also be unexpected (UD) e.g. a phone call or the door bell.*

S- Storage: *It refers to the retrieval or replacement of items, utensils, ingredients etc. from or to an area where they are placed more or less premanently e.g. taking out vegetables from the refrigerator and later replacing them.*

Do	*S	S	S*	*R	R	R*	RF	EA	1	2	3	AO	O	T	D	S	TIME	EQPMT	REMARKS
Do	*S	S	S*	*R	R	R*	RF	EA	1	2	3		O	T	D	S	7.01	-	Takes out Jar containing pulses
Do	*S	S	S*	*R	R	R*	RF	EA	1	2	3		O	T	D	S		-	Picks up plate
Do	*S	S	S*	*R	R	R*	RF	EA	1	2	3		O	T	D	S		-	Places on the counter
Do	*S	S	S*	*R	R	R*	RF	EA	1	2	3		O	T	D	S		-	Takes out pulse from jar
Do	*S	S	S*	*R	R	R*	RF	EA	1	2	3		O	T	D	S	7.02	-	Replaces jar containing pulses
Do	*S	S	S*	*R	R	R*	RF	EA	1	2	3		O	T	D	S	7.03	-	Picks and clob ans pulses
Do	*S	S	S*	*R	R	R*	RF	EA	1	2	3		O	T	D	S		-	Wasnes pulses
Do	*S	S	S*	*R	R	R*	RF	EA	1	2	3		O	T	D	S		Pressure Cooker	Picks up pressure Cooker
Do	*S	S	S*	*R	R	R*	RF	EA	1	2	3		O	T	D	S		-	Puts pulses in the cooker
Do	*S	S	S*	*R	R	R*	RF	EA	1	2	3		O	T	D	S	7.04	-	picks up water jug
Do	*S	S	S*	*R	R	R*	RF	EA	1	2	3		O	T	D	S		-	Adds to pulses
Do	*S	S	S*	*R	R	R*	RF	EA	1	2	3		O	T	D	S		-	places cooker in the range
Do	*S	S	S*	*R	R	R*	RF	EA	1	2	3	Room	O	T	D	S	7.05	-	Answers the phone call
Do	*S	S	S*	*R	R	R*	RF	EA	1	2	3		O	T	D	S	7.06	-	Picks up spiece box
Do	*S	S	S*	*R	R	R*	RF	EA	1	2	3		O	T	D	S		-	Adds spices in the cooker
Do	*S	S	S*	*R	R	R*	RF	EA	1	2	3		O	T	D	S		-	Replaces spice box
Do	*S	S	S*	*R	R	R*	RF	EA	1	2	3		O	T	D	S		-	Closes cooker
Do	*S	S	S*	*R	R	R*	RF	EA	1	2	3		O	T	D	S	7.07	-	Leaves the kitchen
TOTAL													2	13	1	7			

Fig. 8.13. *Recording an Activity on Path Process Chart*

3. Time: The time when the activity starts is noted and the motions that take place within the next one minute are noted next. Subsequently, the motions or every one minute are noted till the entire activity is finished.

4. Equipment: An equipment can be defined as the tool that is used for carrying out the operation e.g. mixer, grinder etc.

5. Remarks: The activity is described briefly in the column e.g. peels vegetables, rolls out 'chappatti' etc.

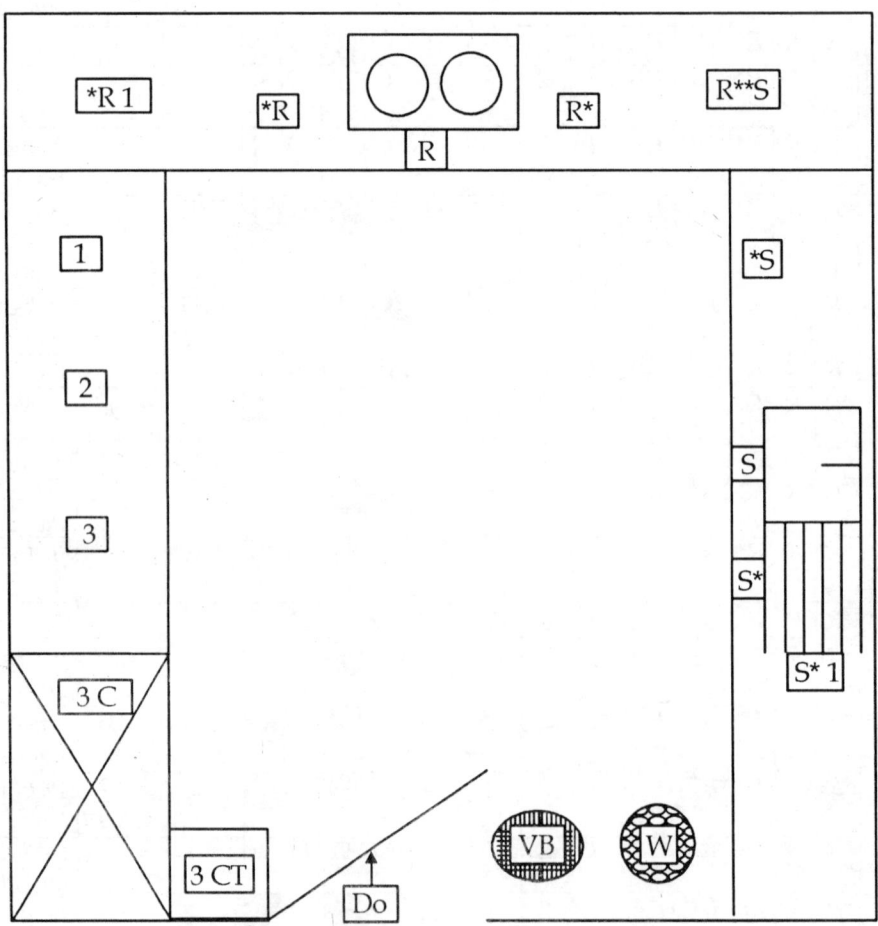

Fig. 8.14. Symbols for the Workplaces: Path Process Chart

Recording the Activity of Meal Preparation

In order to record the activity of meal preparation a layout of th ekitchen is made using a convenient scale (1:25). The placement of the major equipments is also.

In recent years work-simplification techniques have been used by research workers in Home Economics and other fields to improve work methods in homemaking. Motion and time studies have been made of such tasks as food preparation, dishwashing, laundering and ironing, bed-making, cleaning and a number of other tasks. Efficient kitchen arrangement, tools, equipment, storage facilities, and correct heights for work surfaces, chairs, and stools have also been studied.

These studies suggest many methods of simplifying homemaking tasks, which may be used or adapted by homemakers in their own homes. They show how motion and time studies may be applied to any work problem in the home. Trying out new work patterns and adjusting them to meet new situations add interest to the work. Family members are also kept on the lookout for other ways to make tasks easier, to lessen fatigue, and to make work a pleasure.

MANAGEMENT PROCESS AS APPLIED TO ENERGY RESOURCE

Planning Energy Use

Since energy management is directly involved with the performance of work, the major decision in planing this resource is to solve the problems on *when* and *how to perform the work*. It would also require the careful decisions related not only to minimize energy expenditure, but also to avoid bodily discomfort and fatigue.

The problem on deciding *when* to do the work is inter-woven with time planning. This involves the preparation of a time-schedule in which light and heavy tasks are organized alternately. Besides this, there is a need to incorporate rest periods in between these two tasks. However. One should keep in mind that certain tasks are fixed for certain time slots, as in the cases of attending to school or work place, meal times, sleeping periods etc. Considering all these and taking into account one's availability and work demands, an individual should plan her energy use carefully. Becoming aware of one's limitations and the help that can be utilized from others are of greater importance in energy use planning, as it can help her in reducing her work load and the feeling of tiredness and fatigue.

The discussion of time management showed that time and energy planning are inseparable. Managing energy is similar to managing time. It involves the making of activity plans as well as carrying them out, and evaluating the results. In time planning, one's clock or watch helps measure time expenditures, while in energy planning knowledge of the energy costs of different activities, one's skill and ability to turn out work, and the effects of fatigue help to measure energy expenditures. Through experience one learns how to work out well-balanced, energy-spending patterns that are based on the store of energy available from day to day.

Controlling the Energy Plan

Once the work plan in the use of energy is ready, the problem of how to implement it arises. The answer to this lies in the understanding and awareness of certain techniques related to body mechanics, work simplification, etc.

Motivation plays an important role in carrying out all activity plans, including those having to do with energy management. We have seen that high motivation makes more energy available for the tasks at hand, and helps control fatigue costs resulting from work. Motivation can be increased by developing more favourable attitudes toward homemaking, by making work and situations more attractive, and by stimulating greater interest and goal consciousness on the part of the family as a whole.

Developing skill in fitting activities of daily living into the minutes and hours of the day conserves emotional and physical energy and stimulates greater efficiency in work. Skills of this kind lead to greater mental flexibility in the management of energy and in carrying out activity plans.

Work simplification, the effective use of the body in housework, and the skillful performance of homemaking tasks are the real tools which enable a homemaker to conserve time and energy so that more energy will be available for other interests and activities.

Evaluating the Energy Used

Much of the discussion on evaluating the management of one's time also applies to the evaluating of one's energy management. Asking oneself the following questions will help to evaluate success in utilizing one's store of energy:

- Do I think of the use of energy in terms of the goals I wish to attain?
- Do the energy costs of any homemaking tasks seem too high?
- Am I using my energy effectively?
- Have I worked out a well-balanced energy-spending pattern for both my homemaking and my other activities?
- Do I often lengthen my work day in order to finish certain tasks?
- What homemaking activities do I like best?
- What do I dislike? Why do I dislike these activities?
- How can I change my attitudes toward these activities?
- What tasks are most tiring? Why?
- Do I tire easily?
- Do I recognize the type of fatigue I frequently experience?
- Does it make me want to stop working?
- What can I do to relieve fatigue?
- Do I use effective methods of work to avoid fatigue?
- Have I learnt to relax and rest?

An analysis of the answers to above questions would help a home-maker to evaluate whether she is managing her energy resource properly or not.

To summarize, we can say that in order to carry out the work in home, time and energy are the two important resources. They are limited in both quantity and quality, but the study of energy is more complex than that of time. Since the amount of energy possessed by each person is different, while performing the household work, one should be very careful to plan it in such a way so that more work is achieved within the limited amount of available time and energy. Peak loads, work curves, rest periods and work simplification are the tools involved in the management of work time. Out of these, work simplification is the most important tool which not only is helpful in an industry but a very useful tool for a homemaker and more for a gainfully employed home-maker, who has to make use of the simplest, easiest and quickest method for doing work. This can be done by making time plans and time schedules for which certain steps have to be followed. Separate chapter on time management can be referred for understanding and use of time plans.

Besides managing time, the homemaker has to manage energy too. For managing energy, one has to keep body parts aligned, use muscles effectively and also work has to be done rhythmically keeping in mind the centre of gravity and momentum. While managing energy one must become aware and must also understand the types, causes and remedies for fatigue.

Work Simplification-Present Day Scenario

In modern times, medical experts believe that women suffer more from fatigue than men do. The reasons are traced to their biological functioning and reproductive systems. There are socio-psychological reasons too, stemming from the stress women are constantly placed under.

Most Indian men are prone to release stress through smoking and drinking. Few Indian women do this, though more urban Indian women at present smoke and drink too. For them, smoking and drinking are indulged in more as a way of relaxing during leisure than as a stress-buster.

Social factors such as stress and tension are generally linked to personal and family matters, examination anxiety prior to the children's examinations, illness of husband or children and any problem occurring in the parental family etc. These factors affect women much more than the men. Environmental factors such as toxic substances finding their way into our bodies such as pesticide- sprayed farm produce, the use of various insecticides in the homes, the intake of unsafe and unfiltered water, exposure to vehicular smoke and industrial toxic gases and self- medication and the misuse of drugs etc. cause adverse effects on our body ranging from minor skin or allergic problems and fatigue to something as serious as cancer or heart ailments. Environmental and indoor pollution has also reached such heights nowadays that we all automatically suffer from chemical abuse, hence, tiredness and fatigue are felt without any physiological strain. Thus social and environmental factors are more of a concern to be dealt with in solving the problem of stress and fatigue.

Now-a-days the fatigue due to *housework* has been lessened as compared to earlier times because of the advent of labour saving devices such as mixer- grinders, microwaves, washing machines, vacuum cleaners, etc. Also the availability of ready- to -eat packaged foods has considerably reduced the physical labour thereby reducing physiological fatigue.

Healthy Tips for Stress-related Fatigue Management

Recent medical theories and research studies are advocating more physical movement in daily routine such as use of stairs instead of lifts, using desk-top-phones instead of cell-phone or cordless phones, avoiding the use of remote controls for operating television, music system, air conditioners etc. This physical mobility ensures good physical and mental health of the people and promotes healthy practices to ward off the stress experienced by women at home as well as in their work place.

The stress- rich modern life need a different kind of treatment as compared to the measures for work simplification and avoiding fatigue, efforts are, therefore, needed to be taken to avoid stress-related fatigue. Some tips are given below to tackle such situations:

- Adopt a positive mental approach to exercises and diet, and try to make them a part of your daily routine.
- Adopt a healthy life style by incorporating Yoga and Meditation in your routine.

- Involve yourself in recreational activities such as reading, watching T.V. listening to music etc.
- As far as possible try to walk the distance to your home, office, shopping area, instead of travelling by a vehicle.
- Involve yourself in hobbies such as embroidery, sewing, gardening etc.
- Involve in outdoor games such as swimming, tennis, badminton, cricket etc. This can be made to have more fun if the entire family takes part together in these games.
- Avoid eating fat-rich foods as stress-busters. Instead, change the kind of work you are involved in. Take a break like reading, taking a walk, listening to music, talking to a friend or a relative in between strenuous physical work.
- Change the mind-set of the family members by distributing housework among them and ease your workload.
- Group activities in the household should be encouraged by involving children and husband in household work.
- Avoid as far as possible arguments in the house and office affairs. Try to understand that just as you have an opinion, others too have a right to voice their opinion. Talk softly and try to reach a compromise.

Following these tips would not only redure stress-related or bodily-related fatigue, but will also make family living enjoyable and comfortable. These guidelines would provide a long lasting, satisfying and healthy atmosphere for the stress-enriched fast-moving lives in the modern homes.

9

Money Management

A MONG all the resources that are available to the family, the most important one is money. Since majority of consumer goods have to be purchased, or paid for, families are now becoming conscious of the need to learn how to manage their money. They are aware that a better management of limited incomes would provide more of the goods and services they desire.

CONCEPT OF INCOME

Before one approaches the use and management of income, it is necessary for a home maker to understand certain basic facts relating to income and its use. The concept of income is generally not clearly understood because families are accustomed to the use of the term income as synonymous with the amount of money received as remuneration for their services. Money income is the most commonly used term because of the ease with which incomes of families can be compared and their contribution to the economy assessed.

The concept of the family's income in real terms is a recognition of the contribution made by resources other than money to the welfare of the family. Such resources include time, energy and abilities of people as well as resources provided by the community for their use and comfort. When these resources are available and utilised to advantage, money (cash) can be channelized for the purchase of other commodities and services. In this way, the total commodities and services available to the family are increased and provided as real income which is greater than the cash earned. Nickel and Dorsey aptly defined family income as *that stream of money, goods, services and satisfactions that come under the control of the family, to be used by them to satisfy needs and desires and to discharge obligations.* Thus it is clear that income includes not only money in cash, but also other resources like knowledge, energy and skills, services of durable goods owned as well as the advantages families receive from the resources of the community. These resources utilized instead of or in conjunction with money is termed as **real income**.

SOURCES AND TYPES OF FAMILY INCOME

Most families, when asked about the amount of their income, would state their annual/monthly salary or the average weekly wage in terms of only money, although **family income** is that stream of money, goods, services and satisfactions that come under the control of the family, to be used by them to satisfy needs and desires and to discharge obligations. **Money** and **income** are synonymous terms. However, in its real term, money is only one of the aspects of income. It is an important financial aspect with which the families receive their goods and services. It provides the purchasing power to the home maker. It is an indicator of economic and social status of families. As streams of water may be stored behind a dam and sent on through dynamos to generate power, so may this stream of income be controlled and directed to creative ends in a family's living by the process of management.

Before we venture into the study of how to manage money, we should examine the kinds of income that are at a family's disposal. Different authors have given different classification of income. Gross and Crandall has classified the total income of a family into three types which are as follows:

> ➤ Money income
> ➤ Real income, and
> ➤ Psychic income.

Money Income

As stated earlier, this income includes all the income received in the form of money or cash in terms of rupees, like wages or salary, house rent, gifts, interest earned from bank deposits and other investments, insurance claims, and bonuses as shown in figure 9.1. Money income, therefore, includes many more items other than the salary, which is normally considered as the only source of income for the family. Thus, the money income can be defined as the purchasing power of the cash in hand, over a given period of time. This period may be a day, a week, a month, or even a year.

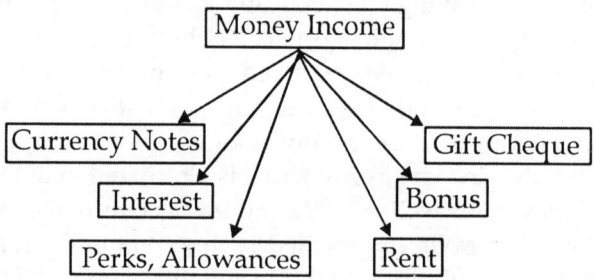

Fig. 9.1: Sources of Money Income

Real Income

Though most families, realise only their money flow as their income, economists include real income too, which is the flow of commodities and services available for the satisfaction of human wants and needs over a given period of time. Real income is derived from owned property and possessions used by family members, such as food grown in a home garden, the use of the house, the automobile and all other equipment and durable goods. Thus it is for the family to realise the significance of real income and manage it well to attain family goals.

At this point, management will be seen to be associated with real income, because the usefulness of certain items will depend upon the use made of them as well as their kind, quality and amount. For example, when a washing machine is not used regularly, or if part of the house is closed off, the potential services of these items will not be fully realized. Thus the real income of a family depends upon the clever use of all items possessed by them. In all, real income is made up of two major types - direct income, and indirect income.

(i) **Direct Income or Non-money Income:** Consists of those material goods and services available to the family members without the use of money. Following are the sources of real direct income:
  - ➢ Commodities such as products of flowers and vegetables gardens and fuel cut from woods owned by the family etc.
  - ➢ Services of all family members including those of the homemaker who cares for family members, prepares food and cleans the house, and those of other family members, such as those of the father who makes car repairs, the son who mows the lawn or the daughter who does family laundry.
  - ➢ The use of a home which is fully owned by the family.
  - ➢ Free or social income provided by the community such as commercial and library facilities, parks, schools, roads, fire and police protection, etc.

(ii) **Indirect Income:** Consists of those material goods and services which are available to the family only after some means of exchange, mainly money. This has already been explained earlier. The commodities and services which are purchased with money range from food items to cars, from the services of a servant maid to the services of a medical specialist.

Occasionally a commodity or a service may be traded or bartered for another. For example, neighbours may give each other some home preparations and students may sometimes trade textbooks at the beginning of a new term.

Psychic Income

The next type of family income is the psychic income which consists of the satisfaction which persons derive from the use of their real income. It is purely subjective and intangible. Although the amount of money income for most families is based mainly on the earning capacity of its members, the accomplishment made out of the money income will depend mainly upon the way it is being used. The quantity of real and psychic income, thus depends largely on the skill that is exercised in the management of their everyday living. Better managed resources will result in a higher accomplishment, thereby leading to higher psychic income. However, differences in psychic income are readily observable. It is closely related to one's standard of living and to the appreciations one has developed. For example, in the case of two families having similar size of composition, income, living and environment conditions, one of them may enjoy higher psychic income by managing their resources better than the other. If one actually possesses the commodities and services he considers essential, he will be satisfied and the psychic income from those commodities, will be high for the individual. On the other hand, the other family whose standard of living is different, may get either higher or lower psychic income from the same commodity or service. For example, one person whose standard of living includes several imported tweed suits receives less psychic income from an additional one than the woman who has never had a garment of tweed before.

Decisions as to the manner in which income shall be used to satisfy the family's needs, desires and responsibilities—collectively and individually-become a major family function. Income management is a family responsibility which can cause considerable tension and much worry, and result in unhappiness if not managed properly.

On the other hand, satisfaction and accomplishment for each member of the family can be realized throughout life if income is managed with thoughtful patience, justice and understanding of the needs of all the members of the family.

Total Income

Total income consists of money income, real income, and psychic income, and in addition, that part of money income which has not been turned into economic goods, and so is not a part of real income. This additional money income may be assigned to three different uses -

➤ Payment of taxes,
➤ Savings for the future, or
➤ Gifts to persons outside the family.

Another classification of family income as given by Betty Stevenson is as follows. He has divided the family income into three types:

1. Productive Income

This is composed of the productive efforts of the members of the family. The productivity of the family members as well as that of the durable goods together contributes towards this income. Thus, this type of income comes mainly from two sources:

➤ household production through the use of talents, skills, abilities etc.
➤ services obtained from durable goods services such as house, T.V, car, stereo, etc.

Thus the efficient use of human and material resources will add to the productive income of the family.

2. Money Income

This is the second type of income. As said earlier, this is the form of money received as wages, salaries, dividends, interest, rent, etc. Thus, it can be in the form of:

➤ Employment income i.e. salary or wages.
➤ Investment income i.e. dividends or interest earned through saving accounts, purchase of stocks and bonds, etc.
➤ Monetary gifts received from members outside the family.

3. Hidden Income

This is that type of income which we generally do not realize. But in a way, this also contributes a lot for the achievement of family goals. This can be of four types:

➤ Community facilities such as police, fire protection, public schools, parks, library, etc.

- Psychic income like satisfaction by efficient use of services and commodities.
- Employment benefits like employee discount, health care, free car, house, etc.
- Consumer savings i.e. saving money through intelligent buying.

Thus the total or the composite income is comprised of the incomes from all sources in addition to the income received in currency. The sum total of family production, actual money earned and hidden incomes represents family's composite income. One should keep in mind that all financial management decisions should be made on the basis of composite income and not merely on the money income. This will lead to better financial management practices, and a higher satisfaction for the family members.

METHODS OF SUPPLEMENTING FAMILY INCOME

Money is not always available to a family in the same quantity. It is different during the different stages of a family life cycle. For a variety of reasons, the family may need to find out the ways and means of obtaining additional income in order to make both ends meet. This need arises if the expenditure exceeds the total of the assured and portable income. Thus, to bring both expenditure and income in a line, it is important that the family should try to find out the different ways to increase the income, rather than foregoing some needs of the family members. Generally additional income is often dependent upon the abilities, skills and talents of the family members. They must be able to find sufficient time and energy to contribute by taking extra works to supplement family income. Following are the different methods by which a family member can supplement the family income:

By Recognition and Proper Utilization of Human Resources: This measure would help a family to save as well as to earn an income. One can supplement the family income by utilizing the human resources like abilities, skills, etc. For example, if a lady is good at stitching, she can start making garments like suits, frocks, etc. on a small scale first to save money by reducing their expenditure on the tailoring charges and then to sell them to earn profit or additional income for the family.

Proper Utilization of Time: As we are aware, time is a resource which is spent without waiting for anyone. So it is important to make proper use of the available time. For example, the homemaker can organise her work in such a manner that she can save time which she can utilize for a part time work or for any small scale or an household production. Thus the time resource can be utilized for productive purposes, and earn additional income for the family's use.

By Household Production and Small Scale Industry: As said above, skills and knowledge can be utilized for the production of goods or may be shared for cooking, sewing, painting, drawing, handicrafts, home gardening etc. rather than employing a servant to do these jobs. Thus the money saved serves as an income earned for the family.

Using Seasonal Food Products: It is always better and cheaper to use seasonal foods. By using seasonal foods, a family can easily reduce its expenditure on food items. One should use very little of preserved and processed foods from the market which are more expensive than the seasonal foods. This can lead to saving of money besides getting better nutrition to its family members. This would also help to use the saved money for fulfilling other needs of the family.

Use of Kitchen Garden: By growing and consuming the vegetables from one's own kitchen garden, one can save a lot of money, as one will not have to spend money to buy them from the market. At the same time. their psychic income will also be increased by the satisfaction gained by the family members, through the efficient use of their time and other resources.

Budget Making: This is an important tool which helps a family in proper utilization of their money resource. By budgeting, the essential needs and wants are satisfied first before the unwanted ones, thereby bringing a control over wasteful expenditure. Budget also has an additional advantage of providing a savings which can be utilized for emergency needs or for buying additional consumable items for the family.

Additional Part Time Income such as part time jobs or overtime work in the office may also be resorted to supplement income. This income also can be termed as wage or salary for part-time or over-time work.

Savings from Family Expenditure is in fact, a way of supplementing income/for, the money thus saved can be diverted to other and better uses, or for buying expensive items. Some times these savings can be accumulated in a bank or any other financial institution to obtain interest or dividends.

Cooperative Purchasing on a Large Scale of vegetables and provisions adds to saving in cost reduction in expenditure on transport and also on saving in time. A number of families can join together and make purchases from wholesale markets which would be cheaper than the retail prices, and thus save money.

Wise Plans for Savings and Investments: It is not always that the savings of the family is deposited in an account in a savings bank. The accumulated savings is further invested into gold jewellery, fixed deposits, shares and bonds or in permanent assets like house, land, etc. Before taking a final decision on the type of investment, a family should find out the right source where the investments yield more profit with security or save taxes. This will help a family to increase its returns from its investments, thereby helping to boost its overall income.

The above are some of the ways and means by which the family can supplement their income with increased satisfaction achieved by the proper use of their resources.

STEPS IN MONEY MANAGEMENT

Income management like other management processes involves planning, controlling and evaluation. The effective management of money involves two types of plans —

- Family's financial plan.
- Family's spending plan.

The family's financial plan includes the entire financing of all needs of the family members, whereas the spending plan is prepared for a specific period. Both these are interrelated to each other. One can't have spending plan without having the family's financial plan and vice versa. The main purpose of these is to achieve the desired quality of life, through effective money management practices.

The main objectives of money management are all round development of the members of the family, enhancing their happiness and health by making the best use of the income of the family. To achieve these goals, the housewife should prepare the family budget which is the first step in money management i.e. planning the use of money.

BUDGETING

While preparing a family budget, an estimated balance is made between the family income from various sources and its distribution on family expenditure on various items for a specified period of time. It can be for a week, month or year depending on the size and tenure of the income and expenditure of the family or according to its budgeting practices. The main purpose of budgeting is to help a family to derive maximum benefit from the family's income for that period. In all, it can be said that a budget is a plan for spending and saving within a given income for a specified period of time.

The family estimates what its income will be during this period and decides how it will use that income in relation to the various wants that the family has. Hence we may describe a budget as a financial pre-plan for future expenditure or it is the statement of results expressed numerically. Budget represents the first step in the management process as applied to money. Its success depends upon its being a realistic, flexible plan suited to the group for whom it was made and also in no small measure, upon the quality of the control and evaluation steps which follows the plan.

Why Budgeting Necessary?

It is now clear that budgeting helps a family to allocate its income to meet its various wants and needs by proper planning. Planning therefore, enables a family to take an overview of their use of income, thus seeing in its perspective. It is a valuable means of comparing various items of income and expenditure.

In addition to this, budgeting facilitates adjusting casual or occasional income to regular expenditure. It encourages conscious decision making which may help in reflecting long term goals in the budget. It forces one to decide what one wants most out of one's life, thereby helping a family to live within its income. It can also help in developing good buymanship or purchasing skills and better consumerism, and it can force a family to save by proper planning. Thus one feels financially secure. A budget gives a clear picture of its various items of expenditure and so helps to identify wasteful expenditure. Adjustments arising from increase or decrease in family size are more easily accomplished when the family operates on a budget. Finally, we can also say that the budget indirectly determines the use of other resources and the kinds of interest which the family is developing,

What are the Limitations of a Budget?

Usually it is thought that budget does not have any limitations, but it should be noted that even the most thoughtfully conceived budget has limitations, such as;

- A budget cannot make an inadequate income adequate. It can, however, help to avoid the use of limited funds for items which the family itself agrees as non-essentials,
- It cannot give an individual the ability to select the products wisely in the market. A budget only directs the expenditure on a particular item.

- Budget making initially is a very tedious process to work out. Only through frequent budgeting practices, one can master the art of budgeting,
- Through budgeting, usually the family with very limited funds surrender all pleasures until the other needs are met, i.e. allowance for recreation is cut off during shortage of income thus making its members to have a feeling of discontentment. Thus budgeting may lead to discontentment to the members of a family with limited income.

STEPS IN MAKING A BUDGET

In reality, a budget would exist if only all the three steps were carried out i.e. first, to estimate income, second, to estimate expenditure and thirdly to bring expenditures and income into line. However, if given only these instructions most people would not know how to proceed, and the resulting budget would probably not be successful. Thus it is admissible to include the following five steps, and these steps require fundamental decisions, and for that reason budgeting is not easy. But a plan should be realistic for the family's use of money and related resources, and it should easily be checked to determine their progress towards their goals. Therefore, the five steps involved in budgeting are listed below:

- The first step is to list commodities and services needed by the family members throughout the proposed budget period.
- The second step involves an approximate estimation of the costs of the desired items, totaling each classification and the budget as a whole.
- In the third step, an estimate of the total expected income, from all sources, of the period in question, is carried out.
- Bringing the expected income and expenditures into balance becomes the fourth step in budgeting.
- In the fifth and final step, plans are checked to see that they have the reasonable chances of attainment.

Step I: Desired Items Determined

This is the first important step of budgeting. In this, a list of all those goods and services which are actually required by a family during the budget period is made.

Since this step involves interpreting the goals of the family in specific goods, considerable thought should be given. For this purpose, some preliminary discussion will no doubt take place among the family members, infact, such discussion may serve as the motivation to budgeting itself; since goals truly accepted and clearly defined are powerful motivating factors for action. The list of commodities and services should have the following three characteristics:

(a) **It should be Classified:** There are a number of goods and services which are consumed by a family over a period of time. It would have no meaning if they were listed without some order. Classification provides this systematic approach. The major headings consist of general groupings such as:

- food
- clothing
- housing and house maintenance
- education

> transportation
> medical
> recreation
> savings
> miscellaneous items, etc.

However, the list needs further classification. For example in listing clothing, the needs of each family member should be listed separately and one may go so far as to classify clothing into items needed for cold and warm weather, clothing bought as materials and then to be stitched, or as ready-to-wear garments, and so on.

(b) **Details Listed:** After classifying the items of a budget under each category, items should be listed in some detail. Omission of some important items at this time may later cause the plan to fail.

Unless details are listed it will be difficult to make accurate estimates as to the amount of money needed or to decide what items (satisfactions) are being eliminated when a category is cut.

After listing the details, now in this step, items have to be arranged in specific order. For example in case of food, it should be further split into—

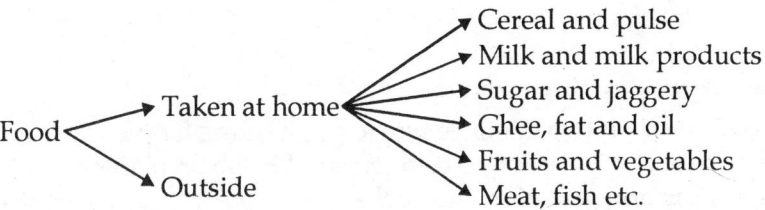

(c) **Relative Importance Established:** It is helpful in listing the items under a budget heading to make some preliminary decisions as to their relative importance. This can be done in either of the two ways:

> grouping the things under "essential" and "nice but not necessary" or
> listing the things "wanted" or "needed in order of decreasing importance."

It is possible that the total cost of the goods needed will be greater than that of the available income. Therefore, an attempt should be made from the beginning to recognise the relative importance of commodities and services which will help in making decisions in the final plans.

Step 2: Costs Accurately Estimated

This is the second step of budget making. If a family is making a budget for the first time, this step will require considerable investigation. Rather it may decide to keep records for a few weeks in order to obtain information about routine costs. As we know that one of the important causes of failure is the underestimation of cost. At every stage of the life cycle or under changing circumstances new demands upon money are likely to occur and the effects of these changes must be calculated. One should also see that general business trends are considered in making these estimates. For example, if prices are showing an upward trend, sufficient allowance should be allowed to cover such increases. One can find out the costs accurately by the following means:

(i) The most satisfactory source of information on costs is the market itself and it is helpful if sufficient time can be allowed during the planning stage to shop around and see what quality is available at what price.

(ii) While the investigation of costs may seem very tedious, this task may be simplified by delegating parts of the investigations to different family members according to their abilities and skills. For example, mother may be a better judge for clothing and textiles while the father can determine the cost of education or buying property.

(iii) Books, magazines and other media can also be used to find the cost.

(iv) Friends, relatives and neighbours can also help in this regard.

Step 3: Expected Income Estimated

After analysing the cost, the next step is to estimate the expected income. It is most important to be realistic in this stage of planning because another frequent cause of budget failures is overestimating income. This can be done by:

(a) Listing income under two headings as possible income and assured income.
(b) All sources of income recognised and then the .income classified under two headings as mentioned above.

Assured Income versus Possible Income

Assured income is the one, which a family is sure to get, whereas for possible income, one is not sure of the time when exactly it is going to be received. When both of these incomes are totaled separately, the family can arrange their plans so that necessities are taken care of out of assured income and the nice but not necessary items can be obtained from possible income. Moreover for families having irregular income, budgeting of this kind is more important.

(a) **Increased Income:** There are many ways and means of supplementing family income which has been discussed earlier in this chapter. If the budget is balanced by increasing the available income, the effect upon the use of other resources and upon the welfare of the family should be taken into consideration, like the effect of women working outside, or a child taking extra coaching classes after the school timings etc.

(b) **Expenditures Cut:** Even if one wants to increase the income but opportunities for increasing the income may not always be available. Thus one should resort to the second method i.e. cutting down the expenditure. Thus the second method of balancing the budget is cutting down expenditure. There are various ways of doing this:

(i) One can cut a little here and a little there by retaining all the items originally included, but reducing the amount allowed for each or the quantity to be purchased. Food is a good example of how small savings can be effected. Many families spend much more on food than is necessary for adequacy, and can make moderate reductions and have equally attractive and satisfying meals with a little thought and care, such as careful shopping of seasonal items only or purchasing at wholesale outlets.

(ii) Another method of cutting down expenditure is by eliminating completely the **nice but not necessary items**. The more irregular the income, the more essential it is that necessities be obtained first and then the optional goods and services be added as more funds become available.

Another method of handling not so regular incomes is to fit expenditures into the income pattern as much as possible. Today, with many families keeping rough records of income for tax purposes most people can estimate what their probable minimum income will be and can use this as a basis for their plans, a wise procedure, since, it is always better to under-estimate rather than to overestimate the income.

All Sources of Income Recognised

In listing income, care should be taken to include all sources such as salary or wages, profits from an owned business, professional fees, bonuses and gifts, dividends from investments and earnings from odd jobs etc.

In addition, since family income is under consideration, income of all members should be included. Thus the total income of all the family members from all the sources during the proposed period should be recognised for balancing the budget.

Step 4: Expenditures and Income Balanced

This is the fourth step in budgeting. This step is of great importance as even the most rudimentary budget is not complete until total expected income equals total expected expenditures. Since needed and desired goods and services ordinarily add up to more than the expected Income, two approaches are possible to bring the two into balance - one can either increase the income or cut expenditures on unnecessary items which are far down the list in order of importance.

Still another method of cutting expenditures can be done by acquiring some of the desired commodities and cervices through direct means rather than through use of money. For example, father can carry lunch to his office instead of eating outside or the family can utilise the products of kitchen garden rather than purchase them from the market,

The budget cannot be considered balanced, however, until the proposed expenditures and savings are mathematically equal to the anticipated income.

Step 5: Plans Checked for Realism

This step, actually a part of planning, is for the specific purpose of ensuring the control of the plan as it is being carried out. The plans, to be realistic, should be checked in the light of the following factors:

(a) **Needs of Family Members Met:** When the balance has been brought about and the budget has assumed a somewhat stable form, it is important to appraise it as a whole to see whether or not it really meets the needs of the family for which it is intended and how realistic it is.

(b) **Emergencies Anticipated:** No matter how thoughtfully a plan is made and how carefully costs are estimated, it is impossible to foresee accurately all demands during a budgeted period. Besides, some unforeseen expenses like accidents, or trip may also arise during the budget period. Therefore, the budget must take care to allow for such emergencies.

(c) **Solvency Assured:** Solvency is the ability to pay bills or debts as they fall due. Even though total annual income may be greater than proposed expenditures, families sometimes commit themselves too heavily in a limited span of time. If a major piece of equipment is purchased on the instalment plan, the family may plan to pay for it in three monthly instalments instead of twelve instalments in order to avoid higher credit costs. Hence, a budget should be able to meet such circumstances too.

(d) **Effects of National and World Conditions Considered:** It is recognised that in modem society no family can live completely into itself, and therefore national and even world wide conditions must be considered in the formulation of a financial plan. The effect of the decisions made by individual families upon the national economy should be recognised, as for example in a willingness to voluntarily reduce spending to absolute needs in inflationary periods or to show confidence by normal spending in deflationary periods. Thus a family might postpone building a house, when the trend of prices is upward or when a war is expected, even though they could afford to do so, at that time.

(e) **Long Term Goals of Family Recognised:** Lastly, a plan is realistic only if the present program fits into and presents itself in the direction of the long term goals of the family. If it is the goal of the family to buy a home in a 5 year period, but they have not been able to save enough toward a down payment in the current year, they must face facts. One of the significant points about major goals is that they are usually difficult to achieve. So, the allocation of resources is required over a long period of time which can be made possible through some allocation in each and every budget.

Whether the budget in its final form is written or not, depends upon the individual family. It is difficult to believe that any family which had thoughtfully carried out all the five steps listed above would be satisfied without a memorandum of the decisions made in the planning stage to be referred to, later. When one considers the infinite variety of decisions which are made in a financial plan for a year, it becomes evident that control of the plan in action is virtually impossible without a written record of the plan for reference. Therefore, the total success of the plan can be made possible only through a well written plan of the financial allocations in respect of the various aspects of the budget. This can easily give the proper opportunity for the family to exercise their control measures while spending, and then to evaluate the success of the plan in future. Besides, the written plan would also act as a reference for making future plans.

CONTROLLING THE USE OF INCOME

In our present day economy, the financing of a family has become a very complicated process. Practically one has to pay for a multitude of expenditures which are made everyday. It is very difficult for anyone to remember all these payments. In order to get an accurate information regarding the expenditure, some system of its recording and other financial operations are needed. This is very important to execute the budget and exercise control intelligently. Even simple records consistently and carefully maintained will help to ease the process of financing for family, and then to control its expenditure.

Similar to the management of other resources, plans are not likely to be carried out successfully unless some control is exercised. This is particularly true with a complex plan such as a budget. Control is as important as the planning. How much one achieves after the successful completion of a budget depends on how much of it is followed. Thus it is very important that proper methods of control are devised.

The process of control in financial management is usually comprised of two areas - firstly checking to see how well the plan is progressing, and secondly, adjusting the plan where necessary.

Checking

Checking is of great importance as it helps to keep in mind the decisions which were made in the planning stage and gives the assurance of knowing whether or not adjustments are needed. There are mainly two types of checks which are followed by many families. These are:

- Mental and mechanical checking, and
- Record keeping.

Mental and Mechanical Checks to Spending

Among the most valuable devices for checking are those which are applied to current expenditures to avoid overspending before it occurs. These may be either mental or mechanical in nature.

Mental checks are usually established by breaking mentally the allocations into units which can be related to actual expenditure. A family may usually check mentally to have an idea about the expenditure that have been done as well those to be done. This method has some edge over the other methods because it is a very simple method without involving the use of a paper and pencil. A person need not have to spend any time by sitting and writing down the expenditures he has made. Many families may also find it easy to follow this method since they may not have to bother about the calculations which they might come across while bringing a balance in their accounts. Thus it becomes a tension-free method. However, a mental check is not a fool-proof method of recording a family's expenditure. It has certain drawbacks too. In this method, the family has no record for future references or for counter checks. Since the family has not maintained any record, at times, it leads to duplication of payments, expenditure etc. This method also has the chances of miscalculations or some items of expenditure may be forgotten. Therefore this method is not recommended for the use of the family since a family expenditure is of multiple and complex nature.

A mechanical check which is frequently used is to set aside a certain amount of money in cash in envelopes to be used for a particular item, and the actual disappearance of the money seems to show how much money is being spent. For example, a family may prepare a few envelopes on the various items of its expenditures such as food, clothing, medicine, etc. The amount of money as allotted in the family's budget is placed in these marked envelopes at the beginning of each month. At the time of expenditure on these items, the money is taken out from the corresponding envelope, and the balance, if any, is put back in the same envelope. The balance of money left in the envelope would indicate the amount of money available for expenditure for the remaining period of the month.

. While comparing this method with the previous one/i.e. the mental check, we can say that it is a better one since there are lesser chances for over-spending on a particular item of the budget. Procuring and retaining the bill inside the envelope or a note about the money spent on the envelope itself can give an idea of how the allotted money was spent. But the family members may make a mistake of spending from the wrong envelopes. The family is also likely to loose it while shopping.

Record Keeping

Record is the second and essential tool in controlling the plan and for keeping an account of total planned expenditure and the actual expenditure made. Thus it helps to keep a financial record of expenditure for future plans. Record keeping is the written record of the exact expenditure made during the implementation period of the financial spending plan. For this, a record sheet is developed for each category of the financial spending plan and in this sheet, the amount allotted and the amount of money spent on these items are recorded. A sample record sheet is given in Figure 9.2.

Item	Amount Allotted	Amount spent in July 2004						Total Amount
	(Rs.)	1	2	3	4	5	... 31	
Milk Vegetables Fruit Bread Eggs Meat								

Fig. 9.2: Sample Record Sheet for Daily Expenditure on Food

The significance of record keeping is made obvious by the advantage it has. The purpose of record keeping is to show the distribution of money which has been spent and to compare the amount spent with the amount allocated. Various studies have been done in this field.

The first study was done by Gross and Zwemer. For the purpose of study, 282 homemakers were taken. Findings showed that only 24.5% families made plans for current expenditure whereas 53.1% maintained an account of their expenditure. The reasons given for keeping records were:

(a) It becomes a proof of payment.
(b) To show what is spent.
(c) To equate what is spent and what is to be spent.
(d) To keep control over future expenditure.

A second study was done by Thorpe and Gross who found contradictory results. They found that a large proportion of people made plans whereas less number of them kept a record.

However, we can list a number of advantages that record keeping provides for a family.

- The record of actual expenditure serve many functions such as the money spent and the corresponding date, the total monthly expenditure, which can be compared with the spending plan throughout its implementation.

- Continuously checking the actual expenditures against the anticipated expenditure clearly indicates as and when the adjustments must be made to avoid excessive expenditure prior to the close of the spending period. It will also provide an evaluating feedback at the termination point of spending time. Record of actual money spent can assist in identifying the areas where the anticipated spending was too high or too low. This becomes a source of input for future spending plan.

- Record keeping provides the data as well as a thorough analysis of actual expenditure. We can also identify our buying weaknesses. There will be indicators of where and how our expenditure reductions can be made. This data will also demonstrate our present spending pattern.

- Records will also point out the possible areas where reduction can be made in spending.

- Unless accurate records are kept separately for actual money expenditure within each category, the data needed for future spending plan would be missing and this information is vital for future planning, and a proof for the successful implementation of the present plan.

Thus record keeping is an important tool used in money management for maintaining and checking the spending practices of the family. The spending plan serves as guidelines for monetary outlay of the family funds, but by controlling this flow by spending and recording according to me established plan, it can be said that the information stored in the records is not only vital to the present plan but also for future financial management.

TYPES OF RECORDS

While a family is maintaining its records, it may adopt any one of the following types:

- Short term records.
- Long term records.

Short term records are those showing the expenditure of realized income in terms of money or cash expenditure, and in terms of real income where goods form an appreciable part of the realized income. These records are called the household accounts and are therefore, the short term records.

Long term records concern items of a permanent nature, such as property owned, other investments, and all indebtedness. These records must be as complete and adequate as possible, since the involved time period is a long one and a family may not be able to remember or recollect all the expenditures it has made during that period.

EVALUATION

This is the third and the last step of money management. After the plan is put into action, it is checked to see whether it has helped to achieve what it was supposed to do, or not. Evaluation of the use of money should be made in terms of the family's goals, to see whether the goals are achieved or any progress being made toward their achievement.

Evaluation of the use of money is particularly important because the satisfaction which the family desires, are purchased with it. In the evaluation stage, the family must not only decide whether planning and their control have helped to achieve the goals which they had set out to attain, but also they must decide whether or not these goals are as satisfying as they had expected. It serves as a feed back to the family. If the family is not able to achieve the goals for which the plan was made, this gives an indication to the family that the goals which they had set out to achieve, is beyond the possibility of fulfilment. It also means that they had not assessed their resources properly or utilised them properly. Therefore, evaluation should be done in the light of specific individual goals, like education to a child, and the family's long term goals like construction of the house. While evaluating its financial management, a family can use a number of devices. There are mainly two devices which help a family to evaluate the results with the original plan.

The first device is **Record or Accounts**. As said earlier, record keeping is an important part which helps to check actual expenditure against the anticipated one. It also provides feedback for future financial planning, thus facilitating the evaluating process.

Besides record of accounts, as a second device **net worth statement** can also be used to evaluate the results. If there is an increase in the net worth it shows that the standard of living of family is increasing and if there is decrease in net worth, it is an indication that standard of living is decreasing.

Thus it is important to evaluate the gains and losses made by a family budget. Evaluation is done not only to assess the success or failure of a planned budget, but also to judge the effectiveness of the control devices applied. This step would help a family to decide whether to follow a similar kind of budget or to make adjustments in the future plan. Since evaluation process critically analyses the budget proposals, it enables changes to be made through adjustments in expenditure proposed to cover emergencies and also efforts to increase the available income for meeting additional expenses.

WHAT IS SAVINGS

Effective money management involves decisions to be made in a manner so as to keep aside certain amount of money from the present income for future consumption. Thus savings and investments form an important component of a family's budget. Savings of small amounts done regularly is always better than a large one done at irregular intervals. The money saved from today's income can be used to fulfill future needs which can either be planned or unplanned. Savings are the result of careful planning. There are various avenues available for investing the money like banks, post office, life insurance, chit funds, provident fund, units of UTI and shares and debentures of public limited companies. There cannot be a set pattern for investing for all families. The families should make a thorough appraisal of all these avenues and then make the decisions to select that avenue which is safe, which gives reasonable rate of return and where easy withdrawal of money is possible.

It was emphasized earlier that while managing money, it is necessary to bring a balance between income and expenditure. While doing so one should not forget that some amount of money is to be kept for future use, so that the family feels financially secure inspite of a fall in the income. This portion of money, which is kept aside from the current income for tomorrow's use is known as savings.

Savings are, thus, the abstinence from present consumption for the sake of future consumption. It refers to the process of keeping some amount from the current income for the purpose of taking care of future needs and wants.

Savings can be defined as the pooling of money either gradually or in a lumpsum in order to meet the future needs. It is possible that these future needs might be planned for unforeseen or such unexpected circumstances, which may be beyond the families' control.

Savings do not accumulate automatically, and in fact, for many they do not accumulate at all. For most individuals and for most families, savings are the result of careful planning and usually the renouncing of some present desires. The motivation for savings may also be the desire for the acquisition of a specific commodity in the future, such as an automobile or a house, which require a large amount of money to purchase.

The desire for some future accomplishment such as college education for a child may also motivate some families to go for savings. A fear of loss of income gives rise to a feeling of insecurity to some families, and this motivates them for sound investments and savings.

To summarise, the following are the reasons for savings in a family:

- Emergencies in the uncertain future, when earnings will not cover needs.

- Retirement, when money is needed after earnings cease.

- Children and family needs, such as for education or for other large requirements, like marriage, etc., and

- Other purposes, such as the need of a house, large equipment and furniture, travel and so forth.

For a family, savings has meaning only when its purpose is well planned and understood by all. Saving for savings sake is futile. Deprivation of the present needs on account of an excessive fear of the unknown future is unsound. The family should know why they are saving so that they may establish a regular plan for saving. The savings programme should fit the family's needs and pockets and be accomplished without undue hardship. An overly ambitious or a poorly planned programme may cause so much tension and discouragement that frustration results and in the end the entire programme of savings will be abandoned.

The significance of savings is recognized the world over, and therefore, 'World Thrift Day' is observed every year on October 30th. Efforts are also made by the government organisations and financial institutions to encourage the families to save more.

ADVANTAGES OF SAVINGS

We can, therefore, conclude that savings play an important role in a family's living conditions. Putting all of them together, we can list the following as the various advantages of savings for a family.

- Savings make the family financially more confident and enables one to face the future bravely without the feeling of insecurity. The main purpose of saving is to provide some portion out of the present income for satisfaction of future wants which would be impossible to satisfy with the future income alone or when there is a fall in income.

- Savings help to bring out a balance between income and expenditure. It is difficult for a family to live from one day to the other without keeping aside certain amount of money as savings for the future.

- Savings help to even out the inequalities caused at different stages of family life cycle. There are ups and downs during the stages of family life cycle. It is advisable for a family to save during the beginning stages when the propensity to consume is less. This portion of income can be utilised during the expanding stages when the demand on income increases.

- Savings help us to develop a systematic layout for spending money. One takes conscious money management decisions if one has savings plan in mind.

- Saving money helps to acquire more money. People include plans for accumulation of savings for investment in business, purchasing of properties, or shares/debentures from any public or private organisations in order to increase the money available. When the accumulated money is invested, it brings in additional income as interest, dividend etc.

- Savings help to even out demands of unforeseen contingencies. Future is unpredictable and insecure. No one knows what would happen in future. Money saved can be used during the future emergencies, both pleasant and unpleasant.

- Savings help to improve one's standard of living by having provision for purchasing a house, an automobile, or any other asset for the family.

- Savings give prestige value to the family in society. One enjoys high status if one has a big bank balance.

- Savings help to face inequalities posed by the business cycle. There are frequent changes in the economy due to changes in business cycle. One should save during deflation when prices are low, and this can be used during inflation times to meet the demands of high costs.

- Savings in the long run at macro level help in growth of the nation. People save money and invest it in organisations which is used for giving loans to business people who return it with interest. This results in the growth of the economy.

SAVINGS AND INVESTMENTS

For the person who is strongly motivated, a decision to place a certain amount with some savings account regularly until the sum is large enough to invest is all that is needed. But for most persons, some method of helping to establish a savings habit is necessary. For these persons, contracting for some investment that can be purchased in small and regular payment units is a more certain method of establishing a savings habit. Such contracted purchases may include chit funds, credit union shares, certain types of insurance policies, savings bonds or other forms of recurring investments.

GUIDELINES OF SOUND INVESTMENT

Before actually investing the saved money, one should have the clear idea as to where to keep the hard earned money so that it is safe and brings good returns. Moreover, hoarding the money at home or in a savings bank account is neither desirable nor practical. Thus one should select the best and most appropriate institution for saving. Some of the criteria for selecting a good investment are discussed below:

  (i) Safety and security of capital
 (ii) Reasonable rate of return or investment yield
(iii) Liquidity or marketability
(iv) Tax efficiency
 (v) After-investment Service
(vi) Time/Period
(vii) Capacity
(viii) Easy accessibility and convenience
 (ix) Effect of world conditions

To summarise, the tests for a good investment avenue are as follows:

➢ Is the investment organisation investor-friendly?
➢ Has the company been in business long enough to have gained a good reputation with respect to earning services and management, and have a higher credit ratings?
➢ Can the nature of enterprise be such that the earnings are not severely affected by changes in business conditions or government policies?
➢ Can the investor wait till the investment fully matures?
➢ Can the security be resold at short notice without sacrificing its value?
➢ Is the saving institution easily accessible and convenient?

AVENUES FOR SAVINGS AND INVESTMENTS

It is important to find out about the various institutions where the consumer-investors can accumulate their savings and invest them accordingly. For an Indian consumer-investor, the following are the facilities available for savings and investment:

- Post Office	- Banks
- Unit Trust of India	- Life Insurance Corporation
- Shares and Debentures	- Chit Fund
- Provident Fund	-Gold, House and Land.

TABLE 9.1: SOME OF THE HIGH RETURN INVESTMENT OPPORTUNITIES.

	Safety	Liquidity	Return
Savings Bank Accounts	High	High	Low
FDs with Banks	High	Moderate/High	Moderate
LIC	High	Low	Low
FI bonds	High	Moderate	Moderate
Govt. PSU bonds, NSC	High	Moderate	Moderate
PF and PPF	High	Low	High
Post Office Savings Certificates	High	Low	Moderate
Company deposits and debentures	Moderate	Low	High
Units of mutual Funds and UTI	Moderate	High	Moderate
Shares	Low	High	High
Chit funds	Low	N.A.	Moderate
Plantation schemes	Low	NIL	Low
Real estate and flats	High	Low	Low
Gold, silver and jewellery	High	High	Low

The Place of Credit in Family Finance

While managing the finances, a family makes all efforts to balance the income and expenditure. But most families find it difficult to maintain the balance between these two. Changes in family needs, income, cost of living and erratic inflation levels force the families with limited income to continuously adjust their expenditure or delete certain items from their budget. Families also make efforts to find out ways and means by which funds could be found to meet this extra expenditure.

In general, families try to keep their income in line with the expenditure in two ways - firstly by setting aside some current income in anticipation of a future expenditure, and secondly by anticipating a future income by the use of credit, or loans from outside sources. If a family is to spend out of savings, it must accumulate funds to become readily available. If it has to make use of credit, someone else must have accumulated capital which this family can borrow to meet their expenditure.

Wise use of credit is mutually advantageous both to the borrower and to the lenders, A mutual benefit or credit funds organisation or a non-banking financial institution is a good example of this kind. In such an organisation some people keep their deposits as savings while those who need funds borrow. By resorting to this method of finance, a borrower can get a home of his own sooner than he could, if he has to save the entire cost of the house first and then buy, besides spending money on rents. The lender also gets the advantage of an extra income from the interest charged from the borrower. The fundamental purpose of credit then is to make it possible for people who require funds for current satisfaction of their wants, to use the funds accumulated or saved by other people with a view to get more in the future for their capital than they would otherwise be able to obtain.

The word credit comes from the Latin word **credo** meaning **I believe**. A merchant will allow a customer to get goods on credit now and pay later, only if he believes that the customer will pay for them as he promises. A lender will make a loan only when he believes that the borrower will repay the money as promised. Any individual's credit is determined by four 'c's, namely character, capacity, capital and collateral.

Character refers to the willingness and determination of the borrower to repay a loan as agreed to, even though it may cost him a great deal and cause more inconveniences than what he would have anticipated. **Capacity** refers to the ability to meet an obligation when it is due, Ordinarily 'capacity' depends upon the income of a borrower. It is important to understand that the capacity of a borrower to repay a loan depends not so much on his income alone, as upon the available margin over and above necessary expenses. The capacity of a family to repay a loan is determined by the savings or the difference between its income and expenditure. The term capital refers to a borrower's net worth, A family's capital is determined by me difference between what it owns and what it owes. The existence of this capital provides a margin of safety for the lender, since the family can draw upon its invested capital when the family's income proves to be inadequate to repay its loan. **Collateral** consists of specific units of capital which are pledged as a security for loan. Usually these units are placed in the possession of the lender, with the understanding that the lender would sell the pledged collateral and recover his debt, if the borrower fails to repay the loan.

Types of Credit

There are two principal ways in which a family may obtain credit. It may use trade credit, with which it buys goods and pay for them at some time in the future. The salesman who sells the goods extends the trade credit to his customers. This kind of credit is found to be very common like the monthly purchases of goods or provisions by the families belonging to lower income groups. Some durable household goods might also be purchased on trade credit basis. The borrower would repay the cost of the goods along with agreed rate of interest on this capital cost over a period of say, every month for a period of one year, two years and so on.

The second way, by which a family may secure credit is to use money directly. It may borrow from an individual or any financial organisation which lends money to needy persons. The family uses this **money credit** to buy the goods it wants and repay the money to the lender at a later date. Interest payable on the amount borrowed may be recovered by the lender either in a lumpsum at the time of borrowing itself or may be recovered separately along with the monthly instalments or recovered in a lumpsum at the end, as may be agreed to between the lender and the borrower.

Whether the credit is obtained directly from the trader or from other sources of financing, the family assumes an obligation to pay a certain sum of money at a given time. When the entire amount of a loan is paid at one time, it is technically known as **lumpsum liquidation**. When the credit money is paid in a series of payments, which is a piecemeal liquidation, it is known as **installment payment**.

Normally the retail stores and commercial banks expect their customers to make a single payment at a given time. However, stores selling goods on instalment basis, and industrial banks and personal finance companies assume that their customers will settle their credit in a series of small payments. Families with high or moderate incomes or families expecting large increments in the future, find it convenient to settle their obligations in a single payment, after receiving their incomes. Families with small, but regular incomes find it more convenient to pay on the instalment plan. However, it costs more to collect obligations in a series of payments than to collect the same amount in a single payment because of the risk of unforeseen circumstances in the future. As a result, instalment payment contracts usually carry a higher rate of interest than a single payment mode.

In many cases, families want to buy things on credit when the need is greatest, so that it can be paid later when its funds are more easily available. Thus families want to borrow money for varying lengths of time. In general, credit granted for a year or less, is considered to be a **short-term credit**. Credit granted for a period from one to five years is called **intermediate credit**. Credit granted for more than five years in considered **long term credit**. Each one of this kind of repayment has its own advantages and limitations. Each family may have to analyse its commitments and future incomes or money inflows to decide upon this. Because of the uncertainties of family life, it is not prudent to assume many obligations running over a long period. When a family is opting for a credit, it is important to know how much it would be paying towards the cost of the merchandise and how much it would be paying as charges for use of the credit. Only after a complete analysis, and also after satisfying himself about the numerous costs involved, the family should avail of the credit facility.

Sources of Consumer Credit

There are a number of agencies which grant credit to families. Different agencies charge different rates of interest, and their terms of lending also vary. A knowledge about these types would serve as a guide to the families while choosing a credit source. The most common ones are:

- Trade credit
- Commercial banks

- Industrial credit banks
- Thrift societies
- Credit Unions
- Mortgage credit.
- Insurance Policy loans
- Pawn brokers
- Personal finance companies
- Unlicensed private lenders

While selecting credit for the family, the family should analyse the merit and limitations of each one of these. For more information on this refer to the book on Consumerism: Strategies and tactics by Seetharaman P, and Sethi, M. by CBS Publishers, New Delhi.

Wise Use of Credit

Before deciding on the choice of a credit, the family must make a thorough study of its availability, the terms and costs involved, and the family's capacity to repay it in the future. Since the use of credit tends to remove some of the usual restrictions upon buying, it is important to be levelheaded and practical in making decisions involving the use of credit. The family must remember that it is not wise to buy credit on a falling market. A family should use credit to anticipate price increases only when the rising market trend is clearly foreseen.

The family with a good character, adequate capacity, plenty of capital, and also has easily saleable collateral is capable of taking a risk with credit and vice versa. If it pays promptly, collection costs are low. The number of installments should also be reduced to the minimum to avoid high return costs. In short, it should borrow as little and for as short a time as possible, to enjoy the credit facility at minimum cost.

10

Space Management ◀◀◀

THE home is a place where we seek refuge from the anxieties and worries of the outside world. It is a place where we enjoy family life through happy group living and where we do the things that help us in developing into mature and emotionally strong men and women. The house is the physical space which provides the atmosphere for the home. For confortable living, the house should furnish adequate space for the daily physical needs as well as for the personal activities of the members of the family. Housing requirements change according to the size, composition and income of the family. An understanding of the part that the house plays in home life and in the changing patterns of family life, will help us in analyzing our housing needs and in utilizing and planning the use of space in our house.

Individual and family values play a major role in converting a house into a home. Some of the universal values sought by all people are beauty, comfort, convenience, good location, health, privacy, safety and economy. All these values are interrelated and influence a family's choice of a house. The manner in which a family furnishes its house and carries on its activities indicates the values which it considers important. Furnishings make our house liveable, convenient and comfortable for carrying on our many and varied activities.

The job of *making a house a home* does not end with furnishing and equipping it. The house, alongwith its furnishings and equipments should be properly cared for if it is to prove functional, useful and satisfying to the family. Sound practices of house care help the members of the family in developing good habits of living and add to the material values of the house and its furnishings. A clean, attractive and comfortable house gives the inhabitants a feeling of pride in their home. A well arranged and well kept house gives the family moral, social and financial advantages that add to their security. The home contributes to the physical, mental and social well-being of the members of the family.

SELECTING A HOUSE

When the amount that the family can afford to spend on housing has been decided, the next consideration is its location. As a basis for comparing various locations, the following factors should be studied in the light of the requirements of the family; the surroundings, the type of houses in the area, the neighborhood, nearness to the place of work, nearness to schools, place of worship, bazaars, railroads and village industries. The house should be located in a place which is free from noise, dust, smoke and heavy traffic. If possible, the locations should give a sense of spaciousness and provide air, sunlight and a pleasant view.

WHAT KIND OF A HOUSE DOES THE FAMILY NEED?

The amount and kind of space needed for housing a particular family is affected by:

- The family income.
- The size and the composition of the family
- The living habits of the family
- The activities carried on in the household.
- The space needed for furniture
- The space needed for storing articles.
- Future housing needs.

The income, size and composition of the family influence the number and size of rooms needed. The number and sex of the children and the number of adults to be accommodated also have a bearing on the number of bedrooms, bathrooms etc. that will be required.

The amount of entertaining that the family does determines the size of the living room, the dining room and the kitchen. Similarly, the recreational activities and hobbies of the family affect the space requirements.

The furniture that a family has or plans to buy will affect the wall and floor space requirements. The position of the doors and windows affects the arrangement of the furniture in the rooms.

The family can determine its needs by making a list of the space it considers necessary for its activities and for its storage.

FACTORS WHICH AFFECT THE CHOICE OF A HOUSE

An important factor in the choice of a house is its attractiveness. The type and size of the house to be selected will also depend on the size of the family, its income and the ages of its members . A family with growing children needs a house with plenty of room to run about and play in. Space is always an important consideration. Here space refers to the space within the compound as well as within the house.

Yet another important factor is the convenience and livability of a house. Good location and the proper arrangement of the rooms in a house make it convenient and livable. Such a house makes house work easy, with the minimum expenditure of time and energy.

Other features of a house that contributes to its convenience are adequate storage space, adequate ventilation and adequate lighting (both artificial and day light). Convenient arrangement of the work areas also increase the livability of a house. The house should provide for the rest, work and recreation of the family.

In many houses, one single room has to cater to all the family activities. In some houses, there may be several rooms, but not enough to accommodate all activities. Therefore, in every house, some rooms must serve as multi-purpose rooms. Flexible room arrangement is therefore necessary for comfort and convenience. This means that the space should be so arranged in the house that it can serve the changing needs of a family.

SPATIAL PLANNING

Planning the space in a house involves the following steps:

- Deciding what activities are to be carried out in a room.
- Eliminating all un-necessary work connected with these activities.
- Providing the conditions necessary for doing the work in the room.

Deciding on the Activities

The first step in planning any room is to decide on the activities to be carried on here. A good way to decide on them is to make a list of the activities meant to be carried on in each room or in each area in the same room.

For deciding on the best use of space, kitchen and its arrangement is dealt in detail in this chapter.

Kitchen Arrangement

The kitchen is said to be the *nerve centre* of a home. Sometimes it is also called the *heart of the home*. In India, a home maker spends about 70 percent of her time in the kitchen . Therefore, the kitchen should be a pleasant and comfortable place.

The kitchen should be such that all the work is done with the fewest possible movements. It should be easy to clean and have good ventilation and drainage facilities. The floor must be firm, smooth and washable.

Kitchen Activities and Work Centers

The usual kitchen activities are cleaning and preparing food, cooking, serving and cleaning up. For convenience and efficiency, a kitchen may be arranged into work centres.

1. Work Space

The kitchen must have ample work space. There should be some space on both sides of the washing area to keep soiled utensils on one side and washed ones on the other. Space is also needed near the washing area for washing and cleaning vegetables, grains etc. There should be work space near the chulha for cutting vegetables, and for making preliminary preparations for cooking foods.

2. Washing Space

There should be a convenient washing area with a good supply of water near the work area. Space should be provided near or under the washing place to keep the broom, the cleaning powder, or liquid cleaner, and the fibre or other cleansing material used while washing the utensils.

3. Storage Space

There should be sufficient storage space in the kitchen for keeping provisions, vegetables, utensils, and cooked food. Storage space can be provided by fitting shelves below the work space in the kitchen, or by constructing racks on the walls. Nowadays, prefabricated readymade kitchen storage is also available which provides more storage space in lesser area and it is easy to clean and maintain.

4. Cooking Space

All cooking processes centre round the chulha and the utensils needed for cooking should be kept near the chulha. Knives and spoons can be hung on a rack close by, salt and other condiments should be placed in a masala box and kept near the chulha so that the homemaker can reach and use them conveniently without wasting much time and energy.

The area near the chulha should have enough light. It should be kept clean and tidy.

Eliminating Un-necessary Activities

Good planning can eliminate unnecessary work. There are many ways of reducing the time and effort spent on house work.

1. Plan a definite place for each activity. Confusion makes work difficult. It can be eliminated by planning a definite place to accommodate each activity. Through careful planning, some work centres can be used for more than one activity. The number of jobs that can be done in one place, without confusion or unnecessary running about, depends upon the length of time it takes for each activity and on the manner in which the work is managed.

2. Arrange work centres in logical sequences. Work centres should be so arranged in relation to each other that the work progresses without any back tracking or delay, in a continuous and uninterrupted path. This reduces the amount of walking and the expenditure of nervous and mental energy

3. Plan work areas that are compact but adequate. Such work areas make work faster and easier, and save energy by reducing stooping and reaching un-necessarily. The work areas must have enough space for the equipment and enough elbow room for the people working on the task.

4. Work areas should be as small as possible, and yet have enough space for efficient work.

   Design work centres in such a way that the materials needed are within easy reach. The following rules help in planning storage space for each work centre:

   (a) Plan enough storage space for supplies and equipment at the work centre.
   (b) Locate storage space within easy reach for the items used most often. Arrange things on shelves fixed at heights, ranging between your finger tips and your shoulder. This renders stooping and reaching unnecessarily,and saves time and energy.
   (c) Arrange supplies and equipment so that it is easy to pick them up. This can be done by designing storage space for particular items to be stored in them.

5. Use labour saving equipment such as cookers, microwaves, mixer grinders, etc. in a kitchen and a vacuum cleaner and a long handle mop for cleaning purposes.

Providing Good Working Conditions

Work areas can be made pleasant, attractive and safe by providing –

- Adequate light.
- Good Temperature and ventilation
- Safety

Adequate Light

Good light lessens fatigue and promotes safety in house work. The source of light, whether artificial or daylight should be so located that it never directly falls on the eyes of the worker and create glare.

Good Temperature and Ventilation

Cross ventilation should be provided in the kitchen to keep it cool and free from undesirable odours. During winter months, a room cooler is desirable.

Safety

Getting overtired is one of the chief causes of accidents in the home. Thus a work area that is planned to save time and energy provides for safety too . Inadequate and faulty equipment, poor light and poor arrangements make a work place unsafe. A safe work area also has free flowing traffic areas/routes.

Housing Norms and Standards

The housing norms and standards have been set up with the purpose of having a planned and ordered construction of housing units so that the dwelling unit has adequate ventilation, sunlight, good sanitation, adequate free space, appropriate drainage with safety and security (Delhi Building Bye Laws 1983).

Minimum Setback

The provision of minimum setback or space for different sizes of plots for all cities. These standards are indicated in Table 11.1.

TABLE 11.1: MINIMUM SETBACK FOR HOUSES*

Plot Size (in sq m)	Front	Minimum set back		
		Rear	Side $(L/R)^1$	Side $(R/L)^2$
Upto 60	0	0	0	0
Above 60 and upto 150	3	0	0	0
150-300	3	3	0	0
300-500	3	3	3	0
500-1000	6	3	3	3
1000-2000	9	3	3	3
2000-4000	9	6	6	6
4000-10,000	15	6	6	6
10,000 and above	15	9	9	9

* Source: Delhi Master Plan 2001 on minimum setbacks.

Minimum Sizes for Rooms in a House

The recommended minimum sizes and space for various rooms in a house are given below:

Habitable Room

No habitable room shall have a floor area of less than 9.5 sq.m. The minimum size of a habitable room for the resident of a single person shall be 7.5 sq.m., the minimum width of a habitable room shall be 2.4 m. and where there are 2 rooms, one shall not be less than 9.5 sq.m. and other 7.5 sq.m.

Kitchen Size

The area of kitchen shall not be less than 4.5 sq. m. with a minimum width of 1.5 m. A kitchen which is also intended to be used as dinning room shall have a floor area not less than 9.5 sq.m. with a minimum width of 2.4m.

Garage

Its size in a residential building shall not be less than 2.75 × 5.4. sq.m.

Pantries

They shall have floor area of not less than 3 sq.m. with the smaller size not less than 1.4.m.

Bathroom and W.C.

The size of bathroom should not be less than 1.8 sq. m. with a minimum width of 1.2m. The minimum size of W.C. shall be 1.1 sq.m. with a min. width of 0.9 m. If it's a combined bathroom and w.c. the minimum area shall be 2.8 sq.m. with minimum side 1.2 m.

Store Room

Size area shall not be less than 3 sq.m.

Choosing and Arranging House Furnishing

Beds, chairs, tables, sofas, curtains, draperies, and floor coverings are all included under furnishings. Furnishings and their arrangement express the personality of the family. By arranging the furnishings in a harmonious design, the house can be made satisfying and attractive. While choosing furnishings the family should consider the amount of money available, the suitability and comfort of the furniture, and the likes and dislikes of the family members.

While designing the interior space in a room, the furnishings should be considered in relation to the size and shape of the room, the colour and design of the walls, floors, and ceilings, the colour and texture of the curtains and draperies, the direction and intensity of the light, and the location of the doors and windows.

The furnishings that are chosen should be useful and comfortable. They should also please the aethetic sense and contribute to the health of the family. The cost and labour involved in the care, upkeep or maintenance of the furnishings should also be considered.

CONTROLLING THE USE OF SPACE IN A HOME

Controlling space use in a home involves consideration of the following aspects:

> ➤ The concept of functional space
> ➤ Storage
> ➤ A well groomd house, hygiene and sanitation.
> ➤ Aesthetic arrangement of space.

The Concept of Functional Space

Work spaces are a major part of the team that is needed to accomplish household tasks. The quality of the design of the work space in terms of the requirements of the task and of the workers has an important effect on the ease with which the work is accomplished . The problem of relating task and worker requirements for design of work space is described by the term *functional design*. It is a design meeting the requirement of the work and the worker. It should enable us to determine and coordinate the requirement of the task and the worker within the available space in a room.

Implication for Designing Work Spaces

The objective of considering the requirements of the worker in designing work spaces is to determine the conditions that result in a minimum of strain on the worker and require a minimum of effort to do the activity. The words strain and effort include not only physical strain and effort but also the affective, cognitive and temporal costs of work.

The objectives of considering the requirements of the task in designing the work spaces are as follows:

- To determine the conditions that will expedite the activity
- To make available the needed space, items and facilities to make it possible for the worker to do the work without allocating much attention to the situation itself
- To avoid the work that does not contribute directly to the output
- To maintain the rhythm of productive activities
- To design the work space to facilitate the performance of the specific task.
- To design the equipment, storage and work surface requirements for a work place according to the nature of the activity but also compatible with the worker's requirements.

The coordination between the *worker* and the *work* helps in creating and evaluating physical arrangement for work.

To develop the physical design of work spaces, a number of aspects must be considered:

- **Location in Vertical Space**
  Height of work surfaces
  Height of shelves,drawers,bins
  Height of appliances or their parts.
  Height and size of items to be stored.
- **Location in Horizontal Space**
  Supplies,tools.
  Adjacent work spaces- height

- **Spatial Arrangement of Parts**
  - Work surface-appliances relationship
  - Distribution of work surface
  - Location of drawers, doors, bins
  - Location of controls.
- **Amount and Dimensions and Space For**
  - Work (length and width of surface, depth of basins)
  - Worker at work—one and more than one
  - Storage of supplies and other items
- **Provision of Special Features**
  - Utilities (electricity,gas, water, drainage)
  - Duplication of facilities
  - Ventilation
  - Lighting

Task and worker requirements must be related to each of these aspects to achieve a functional design.

The Center Concept

The basic components of the work spaces are the work surface, storage, and major appliance or equipment. For many tasks all three components are essential to satisfy the needs of the task and the worker. The bringing together of these parts of a work space to serve a particular function results in a center.

Merits of Centers

Centres save you effort, time and thinking because you have organised your supplies and equipment where you use them first and you have provided storage, work, and holding space for them. Centres reduce the distance travelled in the kitchen, since many trips are made betwen the equipment and the counter and the storage space. If these are within the reach of one another, work is greatly reducd. Work can be more continuous because fewer changes in your place of work and less searching are necessary.

Designing Components of Centers and Work Spaces

The component parts of a center are the work surface, storage and appliance.

Work Surface

The dimensions for the three parts of work surfaces must be planned separately; the height, width, and depth. The total amount of surface needed for the task and the allocation of the surface are related considerations.

Height

The height of the work surface must be keyed to the worker requirement of maintaining good working posture. The height of the elbow is the important landmark for determining the desirable height of work surfaces for house hold tasks, because of the need to maintain a posture and arm position that avoid static work and the fatigue arising from it. For standing work, plan the height of the work surface so that it is 3 inches below the height of the elbow or slightly more depending on the activity.

Width

The width of the surface (side to side measure) is probably more closely related to task needs than to worker needs if a minimum of 18 inches for standing space is provided. Several aspects of the activity affects the width of the space required, the number of items used, their size or the size of the container holding them and what is done to and with the items. Together these factors may increase the space beyond the minimum needed for the worker. Of course, too minimum an amount of work surface will require more work, such as greater attention to find a place to put things, additional motions to rearrange the items, and the subsequent increase in time loss and dissatisfaction with the work space.

The functional limits for the width of work areas range from48 to 58inches for the maximum and 40 inches for the normal work area. (Heiner and McCullough, 1948; Barnes, 1963) . For seated work, the average horizontal distance within reach is less, 39 and 32 inches for the maximum and *comfortable* reaches, respectively.

The maximum work areas identify the distance that can be reached while maintaining a balanced position, the distance represents the side- to- side reach of both arms with finger tips bent so that items can be grasped. The normal work area represents the distance reached without extension of the upper arm, with only the forearm extended. The normal working area represents that area where most of the manipulative action would ideally be done (See Figure 10.1)

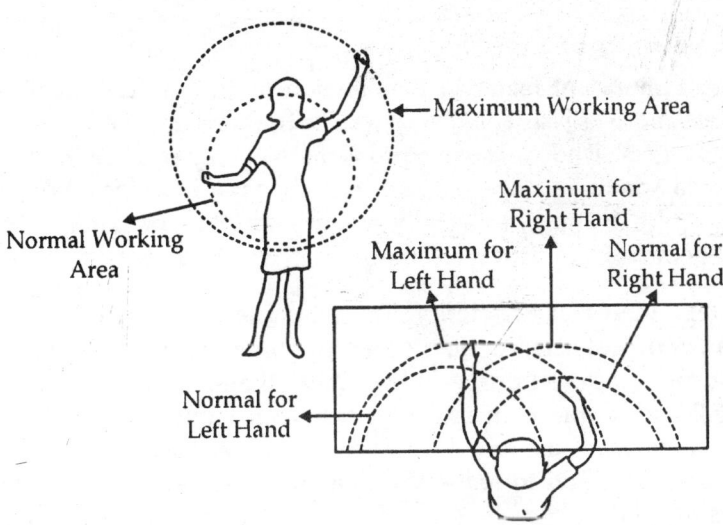

Fig. 10.1: Maximum and Normal Working Areas. (Source: Heiner and Mc Cullough 1948.)

A work surface about 36 inches wide would be within the normal reaching distance of many women and would not require a change of position, particularly when seated for work. The maximum width of about 60 inches would accommodate many women as they work without making them to shift their position. Of course, the shifting of position when standing could have advantages to the work continued over a period of time because shifting position could improve circulation in the lower extremities.

Depth

The depth of the surface (front- to -back measure) is determined by three factors ; the distance that can be reached comfortably, the activity and the items that must be used and stored temporarily during use. The normal working area depth of 14 to 16 inches for women and the maximum work area depth of 24 inches suggest that much of the activity that involves manipulative motions may be done within 14 to 16 inch range but the 24 inch distance is not an excessive depth of the work surface, since it can be reached without assuming an off distance position.

Certain activities require space for free swinging and turning motions. With the maximum work area depths, the worker has more chance of making free swinging motions. Some space for maneuvering both the tools with which she works and the supplies that she uses is advantageous. A compromise between the use of the maximum depth of 24 inches for women and the optimal depth for items that may be stored under the work surface may be necessary.

Amount and Allocation

The total amount of surface needed for work (determined by the width and depth measurements) will vary with the activity. It may also be affected by the allocation of the surface to more than one place. When this is done, a greater amount of surface may be needed because some of the advantage of spacing objects is lost.

Storage and its Advantages

Storage is the most important factor in the creation of an orderly attractive home. *Place for everything and everything in its place* could very well be the home maker's motto. Storage,however, is also the most difficult of all house keeping problems. Not only are even the most modern houses not built to provide adequate storage space, but almost all families, however meager their income, seem to acquire too many things. By working together the family will be able to devise better methods of home storage. This is an enterprise in which everyone must participate.

- **Good Storage Increases Efficiency:** To determine how much time is lost in searching for misplaced or lost articles, note down the time of beginning your task.

  If things are not stored properly then more than half the time is spent in searching for materials and supplies . In a number of homes, many projects have to be abandoned completely because of the loss of necessary articles. The obvious solution is to allot a place for each thing and to request or even force yourself and others to keep it in the right place.

  A convenient grouping of the supplies and materials needed for different jobs is another aspect of good storage. This is particularly important in food preparation. If there is a work center with everything kept within easy reach, house hold tasks would take less time and energy. Household organization often merely calls for planning and imagination for which no extra funds are needed.

- **Good Storage Contributes to Well-ordered Living:** During a search for lost articles, it will be interesting to check the pulse rate . The entire nervous system undergoes a change for the worse when it is found that something is missing. Note the sigh of relief and the lessening of tension when a lost article is found. There are times when the entire family has to join in a hectic search for some articles needed at once.

  Such frantic searches, charged with emotion and tension, cannot contribute to well ordered and serene living. With proper storage facilities, family life can move on with less friction and strain. Time and energy can be released for other worthwhile activities. Confusion and clutter can be reduced or eliminated.

- **Good Storage Conserves Family Resources:** The following situations are familiar enough to most of us:
  A woollen garment is damaged beyond repair by moths.
  Leather shoes are ruined by mildew.
  Milk goes sour.

  The result maybe trifling or disastrous, but there is no doubt that each could have been prevented by proper storage. The material and physical resources of the family are thus presented and protected from the insects, thieves, and such destructive processes as mildew, rust and the like by proper storage. In addition, the general appearance of the home is improved.

- **Good Storage Contributes to Health and Safety:** It sometimes happens that some member of the family falls over a sofa and hurts himself and perhaps even breaks a leg. This is an example of preventable waste of time, funds and resources. More accidents occur in and around the home than in any other place. Cases of children swallowing poison and people getting wounded or killed by firearms or other instruments are further examples of tragedy that can be prevented if these items are stored out of reach, and the heavy household items and furniture are properly stored or arranged.

PRINCIPLES OF GOOD STORAGE

Storage should be such that there is sufficient holding space for all items needed for a group of related tasks done at the same place, the space should permit ready access to each item.

Three principles provide the basic directions for functional storage and then guides are given for implementing the principles.

Principle 1: Store Frequently Used Items at Place of First Use

The first principle requires an analysis of the work done at a particular center to determine what items are used to do the work and their relative frequency of use. The function of the centre provides one key to what should be available. Identification of the needed items guides us in planning the amount of storage space and its characteristics . The second key, frequency of use, provides a criterion when it is not possible to store all the items used first at a centre. Those that are used more often should be given priority since their availability will have a greater effect on the amount of work done or completed.

Organisation of items according to place of first use means that less work is needed . A standing rule for where one item can be found by-passes the need for much thinking and eliminate a search. The preparatory and supportive actions for a task are facilitated. Extra time and effort are not spent in walking to get items. The doing of the task can proceed without undue interruption or delay.

Principle 2: Place Items so They are Easy to See, Reach, Grasp and Replace

This principle favours reduction of mental and physical effort, time and dissatisfaction. Visibility and accessibility are the key factors. Unnecessary costs can mount steadily through repeated use of poor storage. All the factors discussed under storage can be an aid to follow this principle.

Priniciple 3: Determine The Worker's Limit Of Reach

This principle is specifically directed towards keying the physical limitations of the main worker, with respect to her reach, to the design of the storage and placement of items within it .The maximum reach upward helps to determine the height of the top shelf. The level must permit the worker to grasp items from the shelf, but if the worker reaches, across a surface to the shelf, the reaching height is reduced. For these reasons, the distance must be carefully determined.

The worker's normal reach describes the area most suitable for heavy items and most frequently used items if the criterion of minimum input is used. Occasionally we may knowingly choose to place frequently used items outside the normal reach in order that the main worker's use of certain muscles and joints can be increased rather than kept to a minimum or in order to satisfy other family member's preferences.

Guidelines and Tips for Functional Storage

To put the three principles into practice, a number of guidelines are given. In turn, these suggest that the design must be keyed to the dimensions of the items to be stored.

- Sort items to be stored according to the function of the center
- Store unlike items one row deep and one layer deep
- Stack only those items having the same dimensions
- Provide sufficient clearance for grasping and replacing items
- Place frequently used, heavy items within normal reach
- Organize items within the storage space to reduce the search and facilitate the flow of motions
- Group items serving a similar purpose
- Keep together the parts of a piece of equipment when the parts must be assembled before any use
- Be aware of the order of using items or groups of items
- Position items for use to facilitate the flow of motions and eliminate extra ones

The principles and guides for the design and organization of storage are relevant to all activities in the home not only in the kitchen. The numerous examples cited here, from the kitchen are not meant to imply such limited application. Most work and non-work activities require many items and they require storage; laundry supplies, cleaning supplies and equipment, clothing, cosmetics, medicines, bedding and other household linens, gift wrapping and mailing supplies, games, carpenter and repair tools to suggest a few. (Bratton, 1963 and Mc Cullough, 1952). Functional design and organization of storage area for these items facilitate the activities.

A Well Groomed House, Hygiene and Sanitation

A well groomed house is basically clean. There are no odours coming from the kitchen or bath room. No cobwebs are seen clinging to the ceilings or light fittings. Everything is in its place. Windows sparkle, brass gleams, and the furniture reflects the glow of polish. There is a kind of freshness that makes it a home of charm and gracious living.

Hygiene: At any cost all spaces in the home should be clean and well maintained for hygiene purpose, since the health of all family depends on how well the space is kept clean. The following factors are to be considered for this purpose.

Cleaning Equipment

There are tools for every trade and housekeeping is no exception. A basic list of the cleaning equipment is given below:

1. A good step ladder strong enough to take the weight of a person. Ceilings and such fixtures as fans or light fittings cannot be properly reached, and the step ladder is needed then.
2. A long handled brush capable of reaching the most far away spot in the room.
3. Strong brushes or brooms for floors, carpets, mats and rugs.
4. Brushes for upholstered furniture.
5. Special brushes for cleaning the bath room
6. Dusters
7. Mops for wet cleaning
8. Flannel polishing clothes.
9. A scrubbing pail
10. A carpet beater made of cane.
11. A vaccum cleaner

The homemaker should also have the following materials to assist her in thorough cleaning of her home.

1. Household ammonia.
2. Kerosene
3. Strong and mild soaps
4. Disinfectants
5. Cleaning abrasives
6. Metal and silver polishes
7. Furniture waxes and polishes
8. Floor waxes and polishes.

Household Pests

The problem of insect pests is one of the most annoying aspects of the home-maker's job. Unwelcome guests in the form of flies, mosquitoes, cockroaches and other insects visit even the cleanest home. In a clean home, however, they do not generally stay long because of the absence of dirty and damp corners in which alone they can thrive. Insects breed only in dirty places.

A house that is regularly swept and dusted is not necessarily a clean house. Real cleanliness means the absence of the dirt in cracks and crevices or in dark places behind the furnitures. The careful home-maker makes sure that the unseen parts of her home are as clean as those parts which are easily seen. Ridding a home of insects and keeping it free from them calls for vigorous, unrelenting measures and a very high standard of cleanliness.

General Measures for the Control of Insects

The following measures are suggested for controlling insects in the home :

1. Have all doors and windows in the home screened with wire gauze or mesh.
2. Keep on hand some aids for destroying insects, such as an insecticide powder (Gammexane or D.D.T), a sprayer and spraying fluid. Keep a fly swatter handy in every room of the house.
3. Do not permit any pile of trash to remain near the home . Keep the house and its surroundings clean. Weeds must be cut and the path must be swept daily.
4. Fill all cracks and holes in the walls, ceilings, and floors.
5. When living in a flat, pay particular attention to all pipes and openings from the next household. Fill all cracks, whenever possible with a quick-drying cement. Where this cannot be done, keep insecticide powder sprinkled near the openings.
6. Never leave food uncovered. Have metal, glass or other containers with light fitting lids for all staple food stuffs. Never keep food in paper bags, cardboard boxes or other containers easily penetrated by the insects.
7. Regularly sprinkle or spray insecticides into all cupboards and drawers, behind the furniture, in the corners and in all dark places.
8. Regularly air and sun all bedding, mattresses, the upholstered furniture., rugs and heavy clothing.
9. Do not let piles of old magazines and papers in the home. They are not only fire hazards but also harbour insects.
10. Call in an exterminator or fumigate all rooms with sulphur at least once a year.

Insecticides

Kerosene is a good general purpose insecticide to have in the house. When mixed with hot soapy water, it can be used for many purposes, such as washing floors, furnitures, cupboards and the like, spraying etc.

Gammexane and DDT are the trade names of two good non -poisonous insecticides manufactured in India and are easily available.

Fumigation Process

A badly infested house should have the services of a professional exterminator. In many places, the local municipalities provide such services either free of charge or at nominal rates. Usually the job is done in a matter of hours, the home is protected for months. For several days afterwards, dead insects, mice, lizards, etc. can be swept away. A residue of crystals and spots will be left; this will provide protection against recurrence of pests for a long time.

If fumigation has to be done by the home-maker, it is usually accomplished by burning sulphur either in the form of candles, rock or power.

Seal all rooms to be fumigated for eight to 24 hours. Pull furniture away from the walls, making sure that all metal articles are removed from the room. They get tarnished because of the action of sulphur fumes, cover everything with sheets, paper or some other material.

Sulphur is more effective if the atmosphere is damp. The floors and walls should be lightly watered, or containers of hot water placed in the room. Close all doors and windows. Place the sulphur on a metal tray or a lid of an old tin which is supported over a bowl of water. This is necessary to avoid the risk of fire. Pour a little methylated spirit over the sulphur. Light the sulphur, sealing the doors from outside with gummed tape will give better results.

Air the room well before reoccupying it.

SANITATION

Sanitation refers to the control of those factors which affect health. Important among these are water, food, housing, disposal of body waste, and pests like flies, mosquitoes or rats.

Disposal of Body Waste

The disposal of body waste should be done properly. The germs of cholera, typhoid, dysentery and similar diseases as well as hookworms and other intestinal worms continue to breed in the excreta that are deposited on the soil. This waste matter gets washed into wells, streams and tanks and the soiled, germ -ridden water from where it is used for drinking, cooking and bathing. To make the environment and water safe, body waste should be deposited in properly built and maintained latrines.

Disposal of Waste Water

Small pools of water often collect around the bath room, around the kitchen where vegetables, grains, utensils, etc. are cleared, and in the streets. They become breeding places for flies and mosquitoes, small worms can be observed wriggling in such puddles. They are the larvae of mosquitoes and are dangerous to health. Therefore, all waste water should be drained off into a soakage pit, depriving them of breeding places.

Disposal of Animal Waste

Unfortunately, in our country most of the animal waste (dung) is collected, dried and used for fuel. Animal waste is a very good plant food and by using it in that capacity we benefit more.

Many communities have compost pits where dung, leaves and trash are placed. These are then eft to rot and turn into manure.

AESTHETIC ARRANGEMENT OF SPACE

People in India have shown high esteem for any society that has shown sensitivity to beauty in all aspects of its daily living. We admire the people who appreciate and exhibit beautiful things in the areas that permeates into our daily living. Perceptions and the ability to enjoy the expressions of beauty, depend to a large extent, on the degree of sensitivity we developed towards people, ideas and our physical surroundings. The home atmosphere in which we live, is an important factor in determining our values and ideas on beauty. This is the place where we develop our sensitivity towards beautiful objects and surroundings, that results in good taste.

Designing home interiors space is considered as much an art as the other artistic forms like painting, weaving, handicrafts etc. Beautifying the human dwellings is an integral part of everyday living in India, as Indians are firm believers in creating things of enduring value. The varied strands of our culture are woven into the fabric of our day -to- day life. It is no wonder that life in our country is moulded in an aesthetic environment and space. Interior decoration is an art which makes a setting appropriate for a congenial home. It aims at providing comfort, pleasant working conditions, relaxation, a warm and friendly atmosphere along with aesthetic surroundings and factors which provide satisfaction to its inmates, pointed out by Seetharaman, P. and Pannu, P.[1]

The aesthetic aspects of space in a house is dealt in the following ways in this chapter.

- Accessories
- Flower arrangement
- Rangoli.

ACCESSORIES

They are likely to supply the sparkle and liveliness needed to give a drab room a highly imaginative air and their effectiveness bears little relation to their costs. There is no doubt about the importance of accessories in decorating a room. Accessories are the elements that add charm, individuality and vitality to a room. They complete the decorative aspects of a room so that it is truly alive and can be fully appreciated. They play a subtle but a vital role by offering pleasant clues to the personalities of the owners and making the room as unique as the tastes and interests of the people who live there.

Accessories are usually acquired slowly over the years. They may be changed around, or replaced occasionally to provide some variety. Pictures, calendars, lamps, mirrors, wall hangings, statues, collection of books, flower arrangements, clocks, curios, murals, handicrafts, artifacts earthen-ware and ceramic pots and textiles are some of the commonest accessories seen in a home. In the recent times, potted plants are also added as accessories in home interiors.

It is important to mention here that an object meant as an accessory may also serve a function like that of a clock in a living room or a decorative lamp on the bed side. The accessories can either be purely decorative like art objects, antique implements, paintings and so on or be functional like lamps, pillows, mirrors, flower containers, room dividers, bowls and the ubiquitous ashtray. Imagination, inspiration and good taste must go into the selection of accessories . Sometimes a person may accidentally come across an object and know immediately that it is just what she needed for a certain spot in the home. In general, one can say that an item added as an accessory in a room meets all the general objectives of fulfilling beauty, expressiveness and functionalism in decorating a room.

Selection and Arrangement of Accessories

Accessories must suit the overall flow of a room if they are to be effective. The size, shape, colour, theme, texture and purpose of an accessory are to be considered in the first place. Moreover, they

---

[1]Seetharaman, P. and Pannu, P. (2004). "An Introduction to Interior Decoration", CBS Publishers, New Delhi.

should blend harmoniously with the colour scheme that is already established in the room. If the room furnishing are primarily of one period (modern, provincial/American colonial),the accessories should keep to the same period or be in harmony with it. One of the important rule to follow in order to give the accessories their proper setting is to mix formal with formal and informal with informal. A careless mixture never produces successful or the desired results. In all, accessories should meet the basic requirements such as beauty, functionalism, expressiveness, colour, line and form.

Beauty

Most of the accessories are chosen for their beauty alone. It provides the aesthetic pleasure to the members of the family and to those who visit the home. It further reinforces the development of visual tastes and an appreciation of all that is pleasing to the eye.

Function

The homes of today must provide maximum comfort, service and pleasure in return for the minimum care. The accessories chosen may be selected for serving dual purpose of beauty as well as functionalism, for example, books kept in a neat row not only add a pleasing sight but also encourage value for reading and knowledge thereby adding to the intellectual, spiritual as well as the emotional growth of the family members.

Expressiveness

Accessories should express the same idea as the home itself and should harmonize with it. The mood of the furnishing should be extended with the accessory being used. The personality, hobbies, interest should reflect in the selection and arrangement of accessories by the family. Accessories like books, plants, periodicals and hobbies such as painting, stamp-collection, photography, etc. reflect the taste of the family.

Placement of Accessories

The accessories are placed in an interior which has an already existing mood. The accessories thus incorporated should enhance the beauty of the already set trend of the room. The most important accessories in any room should be at the center of interest. The secondary centers should have less important accessories.

One should try to incorporate new ideas to perfect the art of decoration. The advantage of being original is that one can improve and evolve the product constantly unlike those who depend on borrowed ideas. Articles that are placed together should be made of materials that agree with each other. A beautiful piece of sculpture might be placed in the center with a lesser object on each side of the mantel. A large low coffee table may hold a pile of two or three new periodicals, an ashtray, and a plant/cut flowers. A piano should have nothing of these on it (Fig. 10.2).

FLOWER ARRANGEMENT

Using fresh flowers as a decorative accept in a home is an important aspect of decorating a space. There is a growing enthusiasm for this practical as well as visual form that everyone enjoys. It is undoubtedly true that the flowers bring beauty and fresh air in a home and provide a touch of graciousness to the entire room settings. Further, arranging flowers - right from the selection of an arrangement and its corresponding materials in the making of it, to the final touch in the creation of a design of one's own, satisfies the creative impulse of the human beings.

Fig. 10.2: Room Divider, Side Lamp, Indoor Plant, Wall Pictures, Flower Arrangement etc. act as Accessories in a Room

The art of arranging flowers comes as an inborn gift to some. However, it may not be difficult for the others to learn it. Any one who handles and cares for flowers and the other natural things and is prepared to take a little time and effort to collect them, can easily acquire the necessary skills and techniques of this art.

Creating new designs in flower arrangement can be as simple or as complex as one chooses to make it. There is no need to make stiff or elaborate arrangements or to follow a set of rigid rules. For learning and practicing flower arrangements, one should choose an approach namely, the way that leads through one's senses, rather than a purely intellectual approach. Anyone can enjoy using the plant materials in the way they are seen in nature, along with a little bit of one's own imagination.

Steps in Preparing a Flower Arrangement

Any arrangement with flowers must be pleasing to the eyes. When the arrangement is being made, it is possible that one aspect of the arrangement is more satisfactory than the other, or one particular bloom the most beautiful. To arrive at an overall, satisfactory and attractive arrangement, following a systematic step-wise arrangement would be helpful. The steps involved in making arrangement is as follows :

Step.1. Collect all materials required to make an arrangement. The required materials are:

➤ Container
➤ Pin/ Stem holder

> Flowers
> Foliages and fillers
> Water
> Scissors/sharp knife
> Accessories (optional)

Assemble all these materials on a table in on orderly manner. Do not overcrowd the area.

Step 2. Cut the stalks of the flowers to the desired length. This length depends on the type of container, tall or a flat or a round bowl. It is also good to have a fair idea of the kind of arrangement preferably with a sketch, one would like to make. After cutting the stalk of the flowers, put them back in water.

Step 3. After assembling the floral materials, anchor a pin/stem holder firmly into place in the flower bowl/container, using florists clay. Make sure that it stays firmly and can hold the flowers without falling.

Step 4. Start arranging the flowers in the order of their length - the tallest first, the second tallest second and so on. Compare the results with pre-conceived arrangement (Step.2). If necessary re-arrange the flowers. Some buds can be opened partially, or a stem may be bent slightly under the water.

Step 5. Once the flowers are arranged, fill in the spaces between the flowers with leaves or fillers. Use extra greenery to hide the pin holder or the container, if necessary, laying foliage flat at the base of the arrangement.

Step 6. Water is added in the last, after the arrangement is kept in its place, particularly if the container is quite shallow. Add a pinch of salt to the water, sprinkle a little water on the flowers and foliages, if the weather conditions happen to be hot and too dry. If the pin holder is still visible a few pebbles or marble chips can be used to cover it.

To make a finishing to the arrangement, a few accessories, as suggested earlier can be used to make it attractive and appropriate to its location.

Leading Styles in Arranging Flowers

There are three leading styles in flower arrangements- Oriental, Traditional and Modern styles. It would be informative to go into the detailed characteristics of the above mentioned styles.

Traditional Style of Flower Arrangement

Cut flowers have been a source of enjoyment for centuries but originally they were not always put into water but arranged in a garland and strewn over the floors. Gradually they were very often tied together in a bunch and inserted into a vase. The flowers and leaves were arranged in vases of every kind during the Victorian times. During the reign of Victoria, flowers were used very extravagantly in many homes. Decorative pottery-china and glass, stem holders were generally available. At that time flowers from the hot houses and garden were brought together and arranged in showy containers for house decoration in many homes. In large houses the head gardener was generally responsible for arranging the flowers.

Traditional, also called period arrangements are typically full, symmetrical groupings of mixed flowers displayed in a highly decorated ornamental vase resulting in a colourful, expensive looking mass decoration. In such mass arrangements, large blooms and deep hues are usually concentrated at the center; fine flowers or foliage and pale colours are used to outline the basic shape of the design. The overall form of the traditional arrangement may be pyramidal, round or oval depending upon setting and the plant materials.

Modern Style of Flower Arranging

As in painting and sculpture, modern style is a highly individualistic composition. It follows no fixed rules or formulas for correct proportions and employs unorthodox materials for making flower arrangements. They are representative of modern life and exhibit uncluttered, stark lines reflecting on the simplicity of our furniture and architecture and way of living, with which they are more in keeping than the traditional massed arrangement.

Modern arrangements usually are original, made with plant materials minimally but frequently include an unorthodox container as an integral part of the composition. In the modern and abstract styles, artists (flower arrangers) believe in greater freedom in personal expressions. Arrangers working on such styles generally make the fullest use of very striking, dramatic plant forms often in simple yet distinctive containers. To bring added interest for the visual impact, the modern arrangers may leave voids of different size and shapes. The aim is to achieve a strong uncluttered effect with natural plant forms, relying on contrasting textures and shapes for greater impact. Use of accessories may sometimes be essential, not incidental to the design.

Modern and abstract styles are becoming popular because it provides the scope for continuous growth in flower arrangement and embraces new thoughts and methods. Since the style is concerned with personal interpretive expression, their only test is the effect of an arrangement on a sensitive and knowing viewer's eyes. Although arrangements in the modern and abstract styles are original, they owe a lot to Oriental style in concern with their stress on line and a symmetrical balance.

Oriental Style of Flower Arrangement

On examining the pictures of the most ancient art of China or Egypt, it is seen that the flower arranging has been practiced with consummate skill for a great many centuries. In the orient, the art of arranging flowers is centuries old.

The meaning of **Ikebana** is living flowers but more accurate interpretation would be an arrangement of plant material. Nor is it just a pleasing decoration. It conveys simply an interpretation of nature by a pleasing design. It should never be forgotten it derives from a traditional art embracing symbolism and philosophy. Ikebana is the study of nature and art blended into harmony in daily living. It may be understood that this term only came into use in the 15th century and embraces the whole field of Japanese flower arrangement.

Oriental arrangement are more than aesthetic groupings of plant materials. They are symbolic presentations of an ideal harmony which exists between earthly and eternal life. Basic to oriental style is emphasis upon line in every design . In each arrangement there is a triangle. Its tallest line represents heaven; facing and looking to heaven is man; looking to both is earth.

Flowers arranged with artistry are used as constant decorations in Japanese homes. The flower arrangement is an indispensable part of life, and they are displayed in an alcove called

Tkonoma which acts as a frame for the flowers. During the ceremony of serving tea, the guests entering the home are made welcome and then brought to sit and admire the beauty of the flowers for a moment or two of tranquil silence and then the ceremonial serving of tea begins. Some illustrations are depicted in Fig. 10.3.

Fig. 10.3: Some Colourful Arrangements with Fresh Flowers

RANGOLI : THE ART OF FLOOR DECORATION

India has a culture as diverse as her people, climate and size. Indian population is polygenetic and is said to be the melting pot of various races. Inspite of this diversity, it has a rich cultural heritage and is the home for a variety of art and art objects. One such an important art is that of floor decoration, popularly known as rangoli.

Rangoli is an important art originating from India, and is used to decorate mainly the floor, and in some areas, the walls too. The roots to rangoli can be traced to the ancient cave and temple paintings. Though it is a distinctive art of India, it has spread to the other neighbouring Asian countries as well.

Anyone, who visits India, will be treated to the aesthetically attractive painted patterns on the floor, or the courtyard or the frontyard of the houses, specially in the states like Tamilnadu,

Andhra Pradesh, Kerala, Karnataka and Bengal. The luxurious ornamentation of doors, walls and courtyard with rangoli using patterns of singular charm and graceful variety, can be found almost all over the country. One simply has to step into the house of a family during festival or auspicious days to discover the sparkling beauty of this art.

Decorating the floors with patterns is considered as important as cooking food or washing the clothes, since the decoration of the house begins with the drawings of rangoli first in the morning in the frontyard of the house. The frontyard is cleaned and then the drawings or patterns of rangoli is made prior to the start of any other chores of the household. Similarly, rangoli patterns are drawn before the start of any of the religious ceremonies, functions or celebrations. Such is the significance given to rangoli which it is believed, by its presence, heralds all good things in life.

Beautiful and artistic designs of rangoli are, therefore, not only confined to the front courtyard or the place of worship, but also to the verandahs of the house, the centre and corners of the rooms, the main entrance etc. In India, where it is considered auspicious to decorate the house and its entrance, sometimes the entire stretch of the streets are also decorated with rangoli patterns. This is very common during festivals, melas (fairs) and special ceremonies and functions in the temples and religious places.

While making rangoli, the most important factor is the floor. It should be clean, smooth and even. In villages, the floor is given a cow dung finish before making a rangoli. Thus, these floors provide a natural background for making rangoli designs. In the cities, to achieve this effect, many a times, a small section of the floor is specially given a mud and cow dung wash that emulates a typical village -like floor, so that a good contrast of dark and white effect is achieved.

Patterns of Rangoli

The hand-made traditional patterns of rangoli are generally done in white and multi-colored insets and may take the shape of stylized flowers of all kinds, fruits such as mango and almond, animals such as elephant, and fish, bird like the parrot, peacock and swan, and some beautiful and intricate geometrical design with floral insets, all symbolic in form and meaning. Although free hand, patterns are created for beautiful rangoli as of a reputed artist.

To get a thorough understanding of the endless variety of patterns, keeping in mind the traditional and the present and the futuristic aspects of art of rangoli, various patterns of rangoli are classified into six categories.

The patterns used in rangoli can be ritualistic, naturalistic, geometrical, stylised/decorative or a combination of any these. The variety of patterns are therefore, unlimited . It is basically one's imagination which is transformed into a beautiful pattern for Rangoli.

I. Ritualistic Patterns are based on the various rituals, festivals and cultural practices for which our country is famous. Our cultural heritage is deeply associated with various religious sentiments and its appeals have found an expression in the form of Rangoli. Various festivals like Dusshera, Diwali, Karvachauth, Ahoi Raksha Bandhan and the other traditional ceremonies like births and marriages, house warming (grahapravesh), religious practices like fasting and special pujas and havans, are all depicted through artistically sketched Rangolis. In Rajasthan, 'Solah Deepak', (16 Diyas) is made on Diwali. 'Chowk Paglay on Holi, 'Chang' and 'Khera' are made on marriage occasions. Whereas 'Kalash' is for the entrance of the house for happiness and prosperity. Some rangoli patterns are shown in figure 10.4.

(a) ...dry powders for filling

(b) ...on a Theme—"Women's Literacy"

(c) ...with a theme—Helping the Physially Challenged

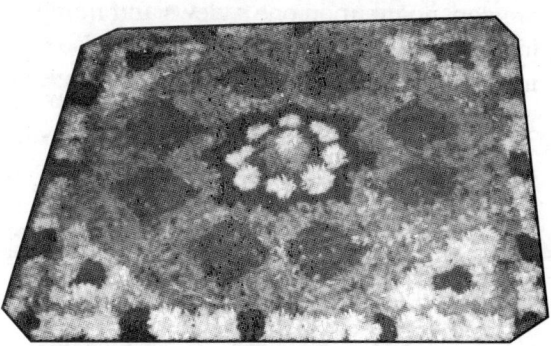

(d) ...Filled with Flowers and Leaves

Fig. 10.4: Rangoli Designs

II. Naturalistic: When patterns are naturalistic in nature, they are based on design of objects naturally occurring in nature like plants and animals and even human beings. Rangoli based on foot prints of a human being is very popular for expressing welcome to guests. Peacock, plants, trees, elephant, fish, kalash, candle, diays (garvi) etc. are some of examples of objects and living beings occurring in nature. In a traditional *tree of life pattern*, a flurry of multicoloured parrots may look vying for honeyed blossoms in green and brown colour.

III Geometrical: Rangoli as an ancient art form largely employs geometrical designs. Most of these designs are made by dots and lines forming decorative patterns. The patterns may be based on line, dot, dash, area, triangle, square, hexagon and circle.

Moreover, geometrical patterns give clarity in the range of design, as well as an idea of clearcut boundaries. All these enhances the beauty of the Rangoli pattern by balancing the use of the available space and colours. For example, on the floor above one inch an area of 20" x 30" can be marked with dots a gap of 2 cms each both horizontally and vertically to get uniformly made dots. When these dots are joined in various ways, unlimited designs can be made.

IV. Plain-line and Filled Rangoli: This kind of rangoli pattern can also be looked into upon as the plain line rangoli as compared to the filled designs. The Alpona in Bengal and Kolam in South are some of the plain line rangoli, which could be geometrical, naturalistic, abstract or ornamental in nature. The filled designs may be filled with rice powder, coloured saw dust, flower petals and grass, mirrors, pebbles etc. so that rangoli design is filled and the floor surface is not exposed at all.

V. Ornamental or Decorative: This is a kind of pattern which is not primarily based on traditional ritual or connected with religion but rather the emphasis is for purely ornamental or decorative purposes. Floral rangolis especially made for welcoming delegates of conference or other official functions in hotel lobbies etc., lend a very beautiful ambience in addition to being very welcoming.

Using bright colours, interesting harmonious accessories, such as statues, mirrors, beads, puja thalis etc. give an ornamental look to this rangoli.

VI. Abstract: Like any other art, rangoli is also practised in abstract form. However, the basic significance is still retained and a symbolic representation of ideas takes place. Although symbols are dependant upon one's views and frame of reference but the ability to symbolize is cultivated in all cultures. Not only artifacts, but also nature's works have always been the objects of man's interpretation. For instance, tree has always been a symbol of life since time immemorial.

Techniques and Materials Used in the Making of Rangoli: The various ways and materials used in the making of rangoli, are listed below:

- Dry rangoli
- Wet rangoli
- Floral rangoli
- Rangoli on thalis
- Rangoli with etching
- Accessorised rangoli.

Dry Rangoli

Dry rangoli was traditionally done with rice powder. However, dry powders of lime, white stone or other coloured powders are also used to make rangoli. It is made mostly in white colour and also with several dry colours. At some places, it may also be done with marble powder or quartz powder. However, locally available materials are also used. For adding colour to the rangoli, either the standard medium i.e. rice is dyed or some naturally occurring mediums are used. Some of the traditional ones are geru- mitti (red mud) coffee powder, neel (blue), roli and haldi (turmeric) charcoal powder, Kalonji seed, sindoor etc. Mica chips or mica powder may also be used to give a sparking effect.

Among the other dry coloured materials, dyed saw dust, holi colours, dyed broken rice, pulses (whole and broken),suji (plain and dyed), henna/mehandi, dyed small marble chips and stones, fruits etc. are popular in the making and filling rangoli designs.

Wet Rangoli

For preparing a wet rangoli, traditionally rice grains are soaked in water for an hour or so, and then ground to get a fine paste. Rice powder in wet form is thus used to make a rangoli. In the place of rice powder paste, chalk powder paste is used for making inexpensive wet rangoli. In modern times, wet poster colours are used increasingly to make this kind of rangoli. These are not only available easily in a large number of shades, but are also easy to paint using a painting brush, for which no extra skill is required.

Normally floral figures or geometrical patterns are drawn using wet rangoli paste. No brush is used, but the patterns are drawn with the finger tips. Though these are generally made in white colour (rice powder), usage of geru -mitti (red mud) paste along with the white lines enhances its attractiveness. Usually the rangoli lines are drawn with the ring finger, which comes with practice. However, a fine brush can be used by those people who have not mastered this art. Though dry rice powder is used everyday, wet rangoli is made on special occasions like festivals, marriages, house-warming etc., in South Indian homes. Special patterns are drawn for *Navgrah and tantrik* pujas for ritualistic purposes.

Floral Rangoli

This kind of rangoli is perhaps the oldest one. Natural materials like flowers and leaves, dry twigs etc., are always the foremost sources of inspiration for human beings for all kinds of arts. It is, therefore, not surprising that natural materials were used initially to fill in the rangoli patterns . Natural materials like flowers not only provide a colourful appearance, but also a fresh and fragrant atmosphere. They are also easy to work with. Any person who is starting to learn to make rangoli, flowers are the best materials to work with. After getting some practice with flowers, one can switch over to other materials like holi colours, coloured saw dust, rice flour etc. Floral rangoli gives very warm and welcoming look. It is easy to make on any type of pattern. Most of the patterns are made using one's own imagination.

Rangoli on Thalis

When people want to make rangoli other than on floors, the best choice is to make it in a *thali*- a brass plate. Traditionally Indians are used to brass plates while keeping various materials like flowers, fruits, during pujas and rituals . Similarly, a portable miniature rangoli can also be made in a *thali*.

Rangoli with Etching

Apart from the traditional rangolis, there is etching work with dry powder . It can be done on the floor, thali or on a platform. The powder is first evenly spread on the surface and the pattern is then etched on this powdered surface. Etching may be done with dry felt pen, spoon, fork, comb, match stick or with the tip of the fingers on white powder.

Unlike for floral rangolis where bigger patterns are used, etched rangoli can be made for small intricate patterns delicately executed, although big sized etched rangolis may also be made as per the demands of the occasion. Etched rangoli may be employed on any kind of pattern from the various patterns discussed earlier. For more details on these areas the book *An Introduction to Interior Decoration* (CBS Publishers & Distributors, New Delhi) authored by Seetharaman P. and Pannu P. (2004) can be referred.

EVALUATING THE USE OF SPACE

A family, while living in a house, constantly evaluates the way it is using the space in a home. As discussed earlier in this chapter, the space in a home is used for a number of purposes. The family has to evaluate the use of space in terms of :

- Adequacy
- Functional areas
- Storage facilities
- Well groomed furnishes and equipment
- Hygiene and sanitation
- Aesthetic appeal.

If the family members are satisfied in all the above mentioned ways, it can be concluded that their planning and control of the use of space in their dwelling unit is proper and the space is well managed. If the family finds that the house is deficient or lacks the requirements for any one of these factors, it has to re-think and find ways to rectify the defects. Since the use of space determines the quality of life a family is going to have, proper space management is important. This is not going to be difficult, if the family is conscious of it and adopts good management practices.

11

Family Life Cycle ⋘⟵

THE first contact for every individual in this world is his family. This individual finds relaxation, opportunity for self expression and happy group living while living in a home as a member of the family. Certain qualities and characteristics are present in each individual since birth and are obviously inherited from his parents. However, some factors like size, composition and income of the family influence and changes the living patterns of the individual. An understanding of the role the family plays in the individual's life and the changing pattern of family life makes it necessary and important for its members to analyze its problems, to find solutions and to make plans for the future.

According to Nickell and Dorsey, besides his mental and physical abilities inherent in an individual, he inherits a natural environment, like the air he breathes and the water, land, vegetation, minerals and animals around him. He is also born into a social heritage, a human and technological environment that is the product of man's efforts and of the human society. This social heritage consists of language, custom, morale, religion, the arts and the sciences, philosophies, social institutions, technological tools and equipments.

Man, along with his heredity, lives in the midst of a three-dimensional environment, namely, natural, human and technological. There is a continuous interaction between the individual and his environment throughout his life. Initially he tries to find ways and means to deal with the effect of the natural forces. In the process of his struggle with the natural process, he creates various tools and methods of various kinds to assist him. This is then gradually developed into a complex technological environment.

With the passage of time, human needs as well as his desires undergo change. New tools are made and new products are developed, and new sources of power are found. New methods are devised to meet these needs and wants. As a result of this unending cycle, new desires arise and with the help of science, man is able to exercise more control over his natural and technological environment, but his human environment becomes more difficult to control. The impact of these changes upon the human being and his social institutions give rise to conflicts, because it becomes easier for men to accept changes in the technological and natural environment than to accept changes in his personal life. He necessarily has to make constant efforts to bring about all these changes together. At this moment, man thinks about **managing** the things to adjust himself to these new changes in his environment.

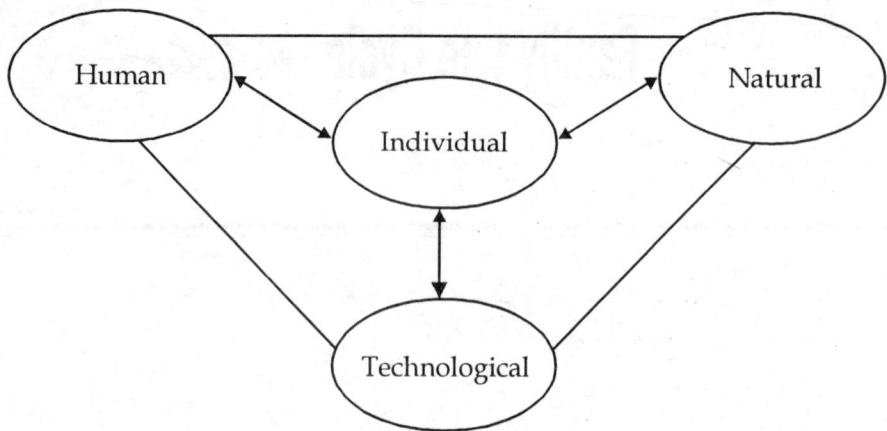

Fig. 11.1: Three Dimensional Environment of the Individual

THE ENVIRONMENTS SURROUNDING THE FAMILY

There are three different environments surrounding the family system. This number is arbitrary, and it might be possible to delineate the other set, of boundaries as well. However, each of the environment described here does have observable boundaries and functions which set off as a separate system, not always compatible with other environments, but always interdependent.

Each smaller system is a sub system of the next larger system. Each environment in turn provides the ever enlarging system in which the family manages. In general, the transactions between the family and its environments — whether input or output — from the view point of the family, can be classified as

   (i) Providing resources or serving as a constraint of appropriate resources, and
  (ii) Either motivating or demanding action and this exchange might take the form of
       effort, material objects, men, money or information.

The Household Environment

The most immediate and the most intimate of the environments surrounding the family is the household environment. It is the area of activity over which the family has the most control and as a result, it most clearly reflects the values of the family. The household environment do not deal merely with the physical limit or the material goods alone. It also deals with the socio-emotional aspects of the individual members of the family.

The Near Environment

The interface between the household and the near environment is somewhat fluid and varies from family to family. The near environment extends beyond the household itself and this is larger and less personal and requires more adaptation on the part of the individual or the family than does the household environment. The near environment in this context includes such groups as business, educational, religious, medical, political and recreational with which the family or its members interact and the facilities these groups provide, and the natural characteristics of the area. However, the character of the near environment is greatly affected by whether it is urban, semi-urban or rural and by its geographical location.

The Larger Environment

The larger environment extends from the near environment to as far as the family interacts. For some uneducated people living in an isolated area, the larger environment of which they are conscious maybe as limited as their state of region. For most Indians, it is their nation. For some others, it may extend even beyond their nation. For some it may be the world and for a small group of people in the recent past, it has included the universe in general, and the moon in specific terms. Some of the families are more aware, some less aware, and some, may not be aware of it at all. However, whether a family is aware more of its larger environment or not, it will definitely have a significant and direct influence upon the family.

THE CONCEPT OF THE FAMILY

The family is a socially recognized unit of people united together by marriage, kinship or legal ties. It consists of two or more people who are economically and emotionally inter -dependant on one-another by residing under the same roof. In a society, the family rather than its individual members functions as the managerial unit which interacts with the other social groups and the environment.

The family is the most important part of a man's environment . As mentioned earlier, the first contact of an individual in his environment is through his family. It is, therefore, important to examine and understand the structure of the family in the modern setting. In order to form a background for the study of management problems, a family encounters in the use of the resources at its disposal, it becomes necessary to understand how a family comes into being, exists and continues to survive. In this context, the family may be defined as a "unit made up of a man and a woman joined in marriage and includes their children." Normally it is in a **nuclear family** that we find the husband and the wife as the parents with their children. These children are the biological off springs although a legally adopted child or children are also sometimes included. The family consists of the parents and their children living under the same roof sharing all their belongings.

In the case of *an extended or u joint family* other members having blood relationships or other relatives through marriage, are also grouped with the members of a nuclear family. These members of the joint family may be belonging to different generations as in the case of grandparents. The extended family can be seen as the merger of several nuclear families. Such a family unit may live either under one roof or in separate houses in close proximity to each other in the same neighborhood.

In a nuclear family, the individual members interact and communicate with each other in close relationship such as husband-wife, father -mother, father-son, mother-daughter, or brother-sister and so on. This family develops a common culture derived both from the general society as well as from the background of the parents joined through marriage. The merging of the culture of the parents together with that developed in the home set up by them gives each family a distinctive cultural pattern. Thus each member of a family can be called a representative of his family when he interacts with other members of a society. Thus the family can be either a nuclear or extended one, on the basis of its structure. On the basis of authority, a family maybe **patriarchal** or **matriarchal**. Under the patriarchal family, the male head of the family is the owner and administrator of the family property and right. The father takes most of the decisions in the management of his home. All other members of the family follow his directions, and thus, are subordinate to him. In the case of a matriarchal family, the authority is given to the woman head of the family and the males are subordinate members. She is the owner of the property and rules over the family. She is the leader of this group. However, in both the family systems, women play a greater role. In a patriarchal family, though the decisions are made by a man, it is the woman, who carries out the activities in the home. In the case of matriarchal family, of course it is the woman, who not only makes the decisions but also puts them into actions. Thus the woman members of the family plays a greater role in the management of home.

Society is made up of a series of functional sub-systems, in which the interchange takes place between the economy, the Governmental authorities, the community and the value systems. All these are received from the family and given back to the family. The economy is viewed as a subsystem associated with the creation and distribution of goods and services valued by the social group. Since management is primarily concerned with the creation and use of resources in daily living, the family 's relation to the economy is of great significance . The acquisition and use of resources by the family will also be affected by the social group. When a family identifies its managerial responsibilities in the house, it is in a position to develop its own conceptual framework of the managerial process. By doing so, the family is able to improve its utilisation level of its resources to enhance the quality of living and thereby increase satisfaction of its members.

ROLE OF THE FAMILY

Family appears to be a general institution among all people. As a family sociologist, Hill remarks I *have been impressed by the universality of the family as an institution in all countries and at all times its great capacity for adaptation and survival.*

It has been seen that the most important part of a man's social environment is the family. Family is perhaps the man's earliest environment where he starts interacting with the other members. As civilisation advanced, the family continued to change, which became more imminent due to industrialization and urbanisation . The family which was self - sufficient in the past, became less self-sufficient and more interdependent, with increased mobility and changes in its ways of earning and living.

One of the prominent changes in the family structure is the emergence of the nuclear family system. The strategic significance of the nuclear family lies in the force with which it connects the family to the larger social group. This is the socialization function. It is the link between the individual and the parts or whole of the social structure. While functioning in a particular sub system of values, members of the family develop and learn to work and cooperate with the others in the larger social group. This is the socialization function. It is the link between the individual and the parts or whole of the social structure. While functioning in a particular subsystem of values, members of the family develop and learn to work and co-operate with others in the larger social groups such as neighbourhood, school, the peer group etc. While doing so, social values are learnt, reacted to, accepted or rejected. Thus the individual changes as the sub-system changes and this in turn affects the system by bringing about further changes therein.

Within the family too, the individual is in a constant sequence of learning situations and during such situations, they also make contacts with some segments of the society. The economy changes with the linkages to the changes in the individual, family and the social structure. In the economy sub system, the family members offer their labour and receive compensation in the form of wages or salary, which in turn is used to buy goods and services provided by the social group.

It is seen in the present day society an emergence of another sub-system. Some of the functions which were earlier carried out by the family and its members are now slowly taken over by the other institutions in the society like the state, educational institutions, commercial or industrial units. For instance, the child care activities that were being formally performed within the family itself at present are done by the commercial or state units like a creche or a day care center run by a play school because of the house wife getting into a full time vocation. In some cases vocation neighbours and friends also come to the rescue of the parents under such circumstances .Thus a non-kinship system emerges to support the family in carrying out some of its activities.

However, the family continues to exist and perform its functions though some changes have taken place in its system.

Every family tries to meet most of the needs of its members by performing certain functions. These can be summed up as —

- Provide the basic needs such as food, shelter and care
- Helps to satisfy economic and emotional needs
- Contributes in child bearing and rearing
- Develops physical and mental security
- Helps to inculcate skills necessary for personality development of an individual
- Helps to establish inter-personal contact, socialization and recreation

Role of Family Members

As indicated earlier not only some of the functions of the family undergo changes with the changes in the society, but the roles of the family members within the group also undergoes changes . With more and more women taking up gainful employment outside the home, her role does not confine only to be being wife and mother, her role also becomes an economic partner. Thus she is in a position to participate in making decision and plans for the family. In some cases, children also contribute to the income of family. These help to extend further contacts and activities outside the family and as such, the family tend to make decisions jointly by the group.

Just as the role of the women in the family changes, the role of the men folk too undergoes some changes. Some of the traditional responsibilities of a mother like care of children, cooking, dish washing, laundering etc. are now shared by the father. Similarly some of the traditional jobs of men such as gardening, marketing, banking etc. are carried out by women, besides contributing to family income through gainful employment. This means that there is no clear demarcation in the jobs carried out by both men and women except child bearing, which has to be carried out by the mother due to anatomical reasons.

Unlike the father, teenage, sons and daughters are relatively inactive as participants in the case of children. However, they may also lend sometimes a helping hand to their parents in some of the day-to-day activities of the household, though some of the tasks are generally assigned according to sex described roles, like the daughter helping her mother in cooking and cleaning, and the sons helping their father in the care and maintenance of the family vehicles etc. Thus it can be said that though there is a general trend in equal sharing of the household responsibilities, distribution of responsibilities in many homes, however, follows the old pattern or sex prescribed roles of the past. However, woman continue to hold the major responsibilities for the household tasks, although a larger number of them have taken up employment outside the home.

Role of Women

There is no doubt that women in the home have been playing a diversified role. Their contribution in the household activities are as follows:

- As home-makers in charge of the operation of their homes
- As directors of household production
- As wives in their personal lives with their husbands
- As mothers in their responsibilities for the development of their children
- As consumer -buyer while choosing and purchasing goods and services
- As court-managers in the use of available resources for the families well- being
- As contributors in the family's income sources

These are the activities which are being performed by woman folk in the family for a long time. Such roles of woman were pre-dominant in the past in an agricultural economy. With industrialization and changes in the attitude and living styles, more woman folk have moved into work situations outside the home too. The direction of their roles has changed, though the difference in the roles of husbands and wives remain more or less the same.

Now-a-days the women have emerged as a

- Colleague of her husband in running the home
- Consumer, or the chooser of the goods when the modern home-maker is supplied with a large number of goods in many forms such as raw, semi-finished or completely ready to consume convenience items
- Wage earner by taking up gainful employment outside the home
- Participant in the community affairs through voluntary social service, religious activities or social functions, etc.

Thus it can be seen that the modern women play diversified roles as a home maker, parent, a wage earner, and as a court manager of family resources in the economy sub system . Among these roles, the most significant is that of a wage earner which is more common now than in the past, when woman confined themselves to the home only. The other significant data is that greater her educational attainments, the better are the chances a women has for employment. There is one more significant data pertaining to women's gainful employment is that of the totally illiterate woman belonging to the economically weaker sections in the society. These women are engaged as workers in the unorganised sectors or as daily wagers as house maids, care takers of the young children, house construction workers etc. Even though a larger number of women are, thus, employed in varies fields, it is seen that as yet they do not have the same level of recognition in the professional areas and in the managerial positions that men have.

Managerial Responsibilities

Management of resources in the family living, mainly centers around the following duties assumed in some degrees by all men and women in their homes:

- Developing a sound philosophy with sustaining values
- Setting family goals
- Establishing satisfactory relationships
- Planning the use of family income
- Planning the use of human resources such as time and energy
- Developing abilities and skills
- Acquiring diverse knowledge
- Guiding educational and social development of family members
- Choosing suitable housing
- Choosing goods and services needed by the family

An analysis of these managerial responsibilities shows many relationships. Some responsibilities are the direct outgrowth of the others, and as such are conditioned by them. Family finances and their management, for instance, touch all phases of family life, because they influence desires, decisions and choices, and in a large measure, control what can be obtained or done within the available resources . In planning for clothing, one must first consider the family's financial resources and decide on the most suitable ways to fulfil the desires and needs with the money available. The management of time and work is also closely related to the other management responsibilities. In order to accomplish each day's work without undue stress and tension, a homemaker must constantly keep in mind all managerial responsibilities and their related tasks in terms of time and the effort required for the various aspects of the living .

It is of significance here to point out that only through careful study of the mental and physical processes that come into play in meeting managerial responsibilities in the home, can we get a clear understanding of what is required, with all its potential problems and obstacles, and its many and deep satisfactions. Thoughtful analysis and decision making in the earlier stages of home making are the bases for sound decisions, as additional responsibilities are assumed. Since all these revolve and resolve around the use of family resource, family participation in these decisions is essential for successful management of the modern home.

CONCEPT OF FAMILY LIFE CYCLE

Each family passes through a cycle that begins with the marriage of two young persons, grows with the arrival of children and then again becomes a home of two persons, when the children marry and leave their homes. Thus the family starts with two relatively young persons, grows normally into a large group of assorted ages and finally returns to a group of two older persons. Though there may be some differences between the families, each family goes through similar stages in their life history. These stages tend to overlap each other in most families. For example, in a family where there are more than two children, one child might continue to grow with new arrivals and the eldest child might get married and leave the home. In this manner, this family expands and contracts at the same time. However, the family faces clearly defined situations along with the problems typical to that stage.

IMPORTANCE OF RECOGNIZING THE FAMILY LIFE CYCLE

Understanding the concept of family life cycle is a valuable guide and in realizing and meeting the managerial problems of family life for a young couple who constitute a family after marriage . The importance of recognizing family life cycle of the family lies in the opportunity it provides to the family of two young inexperienced couple to look ahead and to foresee the family needs and wants which may be expected to arise during each stage. It will also enable the family to predict the related problems and the flow of resources and funds it can reasonably expect at various stages. Many of the problems which a family may face at a later stage may be due to the fact that it was not prepared adequately to meet such situations either due to lack of knowledge or on account of circumstances beyond their control. If any provision was made in the earlier stages of its life cycle, it would have certainly enabled the family to be prepared to face such situations without any difficulty.

Among the various problems that most families face and the most important of them, is that of finances. However, demands on other resources such as time, energy, abilities, skills and material goods would also be felt as the family grows with the arrival of children . The various demands of the family life cycle is depicted in Fig. 11.2.

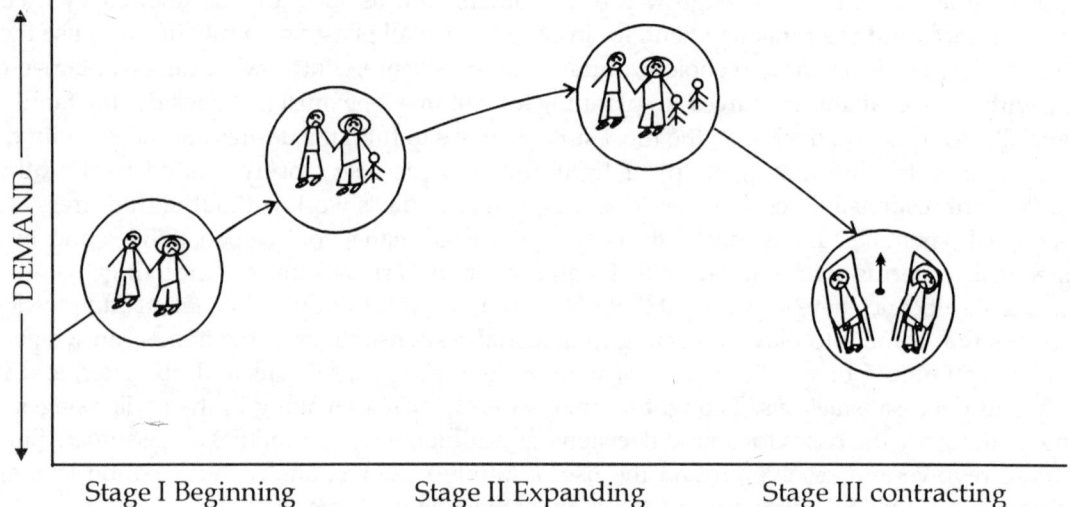

Stage I Beginning   Stage II Expanding   Stage III contracting

Fig. 11.2: Stages of Family Life Cycle, and their Demands

As is evident from the figure, it is important that the young couple realise the extra demands that are likely to arise after the arrival of the children. Though in the beginning, their personal income and the other resources available may be found to be adequate, an understanding of the concept of the family life cycle would enable them to reserve or set aside a portion of it for meeting the additional demands which are likely to arise in the future during the second stage. Thus it enables the family to plan for the future at a time when the flow of resources may be found to be inadequate as compared to the demands arising with the arrival of the children in the family. However, the income might once again become adequate when the children grow and take up gainful employment and contribute to the family's finances, or become independent and leave the home. At the end of the life cycle, reduced earning power of the bread winner may lead to inadequate earnings. Such may be the case when he or she retires from service either with no income or a pension which may be much below the actual salary of the person. Figure 11.3 indicates the flow of resources and the demands of the family at various ages of the bread winner.

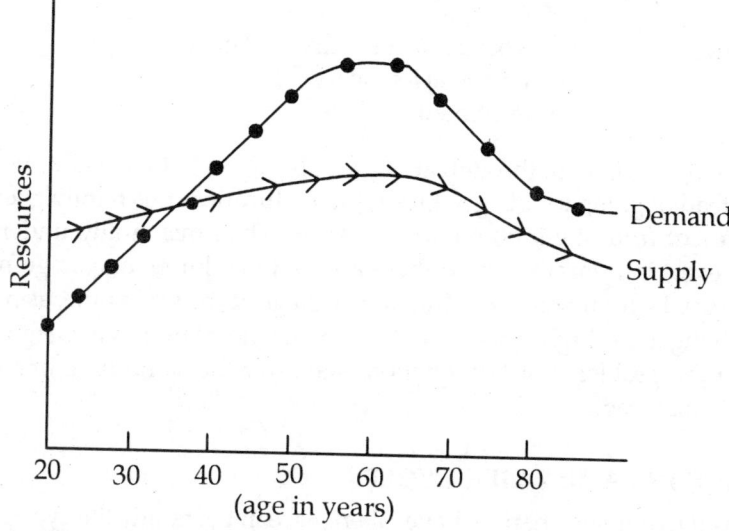

Fig. 11.3: The Resource Cycle — Flow of and Demand on Resources

In India, normally an individual completes her or his school education at the age of 16 to 18 years, and finishes graduation around the age of 20 years. Most prefer to take up vocations or gainful employment at this stage though some may opt for higher education. Once settled in a vocation, they may also marry between the ages 20 and 30 years. If we assume that a family is founded at this stage, as can be seen from figure 11.3, the availability of resources remains more or less the same till the age of 60 years when one of the family members retires from service. The income flow is reduced alongwith the human resources like time, energy, etc. with the advancement of age. Once the children are born around the age of 30 years, the demand on resources also starts increasing till the age of 50 years or so, when the children would have grown and started contributing to the resources of the family. Thus the supply and demands of resources come to a level where both are more or less equal.

ANALYSIS OF FAMILY LIFE CYCLE

In general the family life cycle can be divided as suggested by Gross and Crandall into three major stages, usually called -

- The beginning family - Stage I
- The expanding family - Stage II
- The contracting family - Stage III

Nickell and Dorsey also suggest similar division of family life cycle. Bigelow further added eight sub-stages under these three stages. These sub-stages are the following:

Family Stage	Sub-Stage
I. Beginning	1. Period of establishment
II. Expanding	2. Child bearing and pre-school

III. Contracting

3. Elementary School
4. High School
5. College
6. Vocational adjustment of children
7. Financial Recovery
8. Retirement

Most families with children go through all these sub-stages. If, in any family, the children take up a vocation instead of going to college or for higher education, this family goes to the period of vocational adjustment from the high school sub-stage. Thus this family undergoes only seven sub-stages. However, in a family where there are more children, one stage may overlap into another or, there may be more than one child in one stage at the same time, as may be the case of two children studying in the high school or college. Since most families have more than one child, they have to face the problems of two or more stages at the same time or to face intensified demands in a particular stage.

RECENT TRENDS IN FAMILY LIFE CYCLE

It has been observed that many changes have taken place in the family life cycle in the recent past. It has been seen that both the typical length of the three major stages in the cycle and the length of the entire cycle varies significantly .The reason for these changes are:

- Emergence of nuclear families
- Choice for a small family
- Increased longevity of family members due to improved medical facilities and an enhanced awareness among the people
- Late marriages on account of emphasis on women education and employment

Due to these reasons, the possible changes that arise in family life cycle are as follows:

- The child bearing stage becomes somewhat shorter
- The marriage of the last child comes later
- The wife and the husband have a longer time together after the marriage of their children

In India, as a tradition, children were married at an early age even before they attained adulthood. It was also seen that the parents were taken care of by their children . However, in the recent years, a different trend has set in the families. The children leave the homes of their parents and set up homes independently .With the result, the retired parents are left alone in their homes to take care of themselves. In the cases where one partner dies, the other partner, who is left alone sometimes has to take shelter in the old age homes. Thus, the last stage, i.e. the contracting stage, becomes longer and there is a need to open up more old age homes, and the couple is also expected to save enough to meet the demands of this long period.

Thus it can be seen that a knowledge of the life cycle of a family and its changing patterns helps a family in planning for the future. It enables them to look ahead and picture their needs and desires. The old phrase *saving for a rainy day* has real meaning when one understands the family life cycle.

IMPACT OF RECENT CHANGES IN THE FAMILY SYSTEM ON THE HOUSEHOLDS

Changes in the larger system influence the quality of life in the household. Families are now more inter-dependant and less self-sufficient than when each household produced and consumed the majority of its own food, clothing, housing and their relative services.

Industrialisation and the trend towards a service economy have produced four major changes for the individuals and their families -

- Dispersion
- Mobility
- Producer-consumer
- Small family

Dispersion

Family members shared work on small farms or in business in the past. The hours of work might have been long, but the entire family shared in responsibilities in the production process. Nowadays wage earners may not spend as many hours in earning a living as they did before, as this time is normally spent away from the living unit. The wage earner's occupation, in large, determines the status of the dependent family members. On the other hand, young family members have many job related skills away from the living unit. This dispersion of family members make time for them in developing inter-personal relationships outside their homes. However, inter-personal relationships within in the family group has become difficult to arrange. Companionship among family members is now more related to leisure than to work.

Mobility

Families in the yesteryears were confined mostly to their home towns. As producers of their own consumption products, they stayed in one place. There was less of mobility in physical terms. However, with industrialisation and on account of other social changes, families started to move from one place to another. Families usually move to adjust themselves to the demands and requirements for space, location, employment, design of their living unit or related needs and desires. Most of the family members changed their residence either at the time of marriage or within a year after marriage . However, the mobility rate was gradual.

Producer-consumer

Families have become consumers rather than only material producers. Products can be purchased in a ready to use forms for immediate consumption or in a partially prepared convenience forms to be completed as and when needs arise. Use of currency trade in the place of barter system has facilitated the process of the producer consumer to become just a consumer. In addition to the changes of this kind in the market scenario, the services available to the family have also increased. This has lead to an overall change in the family system.

Although a majority of the families enjoy the changes of affluence, low income consumers spend the majority of the funds for necessities of food, housing and clothing. Not all the production has left the living unit. As consumers, family members continue to transport goods. Some home production activities are continued because of the feeling of satisfaction derived from **do it yourself** projects.

Small Families

The joint family system flourished in the days of the past when agriculture and trade in the village were in a sound position. But, today with the establishment of new factories in urban areas, the families started moving towards cities where they have become economically independent. With the influence of western culture and deficiency of space in the houses, the joint families have disintegrated and small nuclear families are the norms of the present day living. Technological advancements have also changed the living styles of the families. In small families, there are fewer hands to perform the household chores. Modern technology has also changed the family organisation, relations and functions in many ways.

Unlike joint families, the men folk in small families started playing greater roles. The typical roles of women in the areas such as cooking, cleaning, washing, child rearing etc. are now being shared by male members of the family. These roles have become more crucial when the women are gainfully employed. Women, therefore, started contributing to the money resources for the family, and thus, are economically independent .They participate while making major decisions of the family. Many of the household productions have been replaced by the market products . Ready to use consumer and convenience foods, fast foods have dominated the household means . Laundering and ironing, cleaning and many similar activities are no more performed within the house. They are performed either by hired workers or done by agencies outside the home. The joint families had many members to share these activities and in small families, there is a shortage of man power. Therefore, the emergence of small families has changed the management system in the families.

SMALL FAMILY VS. LARGE FAMILY

With the overall changes happening to families in the modern days, the family system has also undergone changes. More families are adopting small families to balance the demands on their resources. Some of the characteristics of large families that are perceived by small families are:

- The per capita availability of resources is limited.
- There is an overlapping of stages and sub stages of family life cycle.
- The parents of the large families find it difficult to channelize the energy of their children
- These parents are not in a position to look after the emotional and psychological needs of each and every child.
- There are more of such situations when conflicts and rivalry arise among the children.
- The eldest girl child is mostly given the responsibilities and is expected to carry out the role of a mother.
- There is generally no democracy in a larger family and chances for a single authority is high.
- Because of financial problems, the launching of the children becomes difficult, and therefore, not good.
- There is no contracting stage for the parents.

RECENT CHANGES IN FAMILY LIFE

Families respond with resilience to changing social conditions. Rapid industrialisation throughout the twentieth century has been accompanied by the following changes in family life:

- More men and women get married at a later age.
- More families have two or less than two children.
- More persons live to complete their family life cycle.
- More women work outside the home.
- Families have moved off farms/rural areas and into cities and suburbs.
- Families have shifted from production to consumption
- Families have more resources, specially the material ones
- Family members have more leisure and higher education.
- Family roles have changed.
- Family instability has increased.
- Individual family members have more freedom.

The image of family life continues to change as families reflect social change. Now, freed from traditional drudgeries, today's families are more specialised than ever before as they focus upon the development of their members as their central goal .

MANAGEMENT OF RESOURCES DURING FAMILY LIFE CYCLE

The analysis of family life cycle reveal that the family has to pass a number of major sub-stages which are likely to have their own and typical demands on family resources. Therefore, it becomes necessary to study the chief characteristics and problems of each of these stages of family life cycle. It is also important that the family understands the factors that influence management and changes in the availability and use of resources during its course. Among the various resources available to a family, the significance of some important resources like money, time and energy are felt more than the others, as their availability affect and determine the availability of the rest. Therefore, management of money, time and energy resources are discussed earlier than the discussion on the different stages of family life cycle.

Money

Money is an important family resource for its significant role in family resource management. Money resource, otherwise known as income, starts relatively low in the young family, becomes larger in the later years. The increase in income reaches and maintains its greatest level for as long

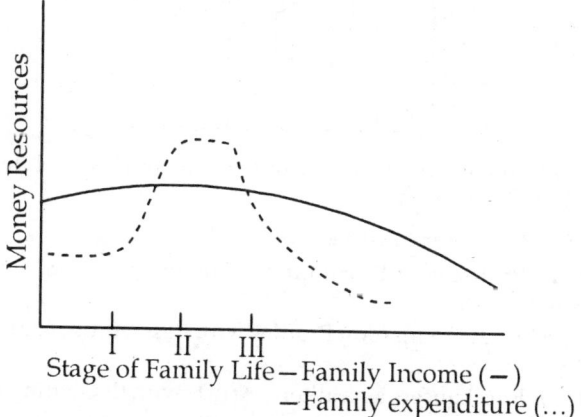

Fig. 11.4: Income and Expenditure Patterns During Family Life Cycle

as the husband and the wife are in employment. It starts dropping with the retirement of one of them and dwindles further when the second member also retires. A possible graph showing the flow of the income and its use during the life cycle of most families is shown in figure 11.4.

For most families, the possible income and expenditure during family life cycle will remain more or less the same as shown in figure 11.4. But the flow of income will vary depending upon the kind of occupation family members opts. Nickell and Dorsey have described life time income profiles of some occupational groups; which in turn would help to visualise the flow of income for these groups as shown in figure 11.5.

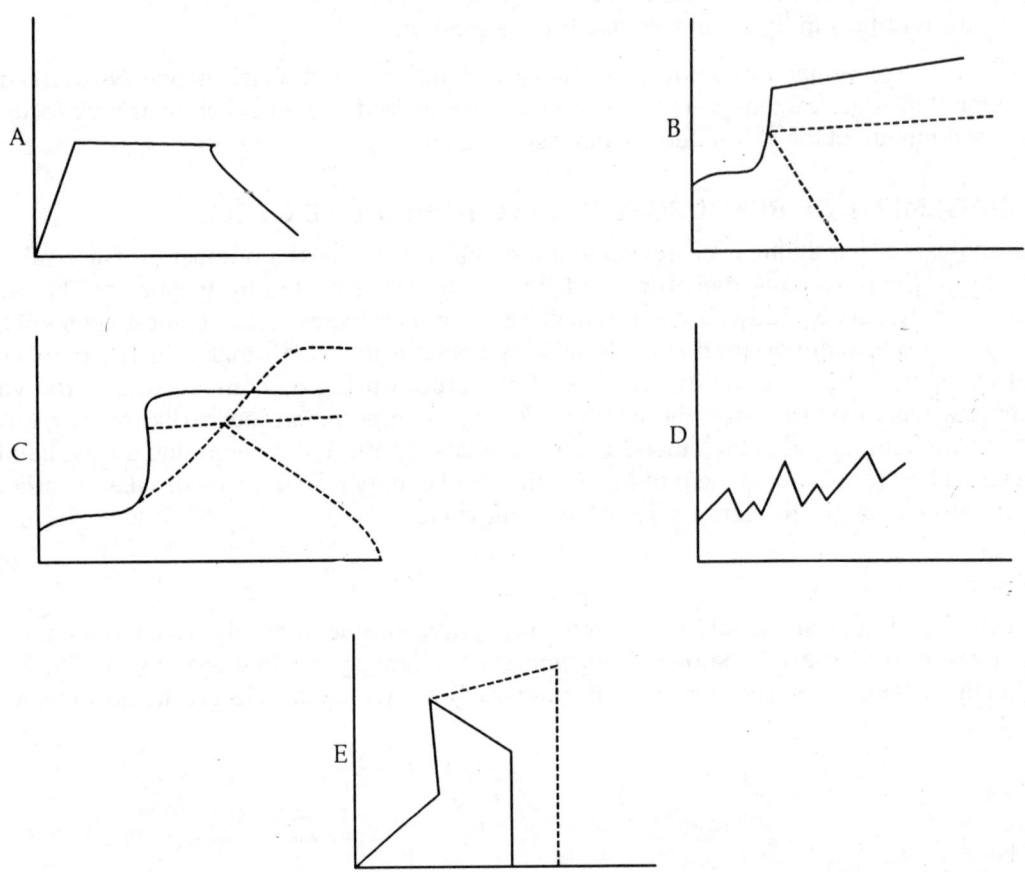

Fig. 11.5: Profiles of Family Income During Various Stages of Family Life-cycle for Families with Different Kind of Occupations

A. Medium Income Profile for Most Families

B. Inherited Income Profile with Possibilities of Increase (−) or Decrease (--) in Maintaining Principal Amount

C. Professional/Technical Occupation Profiles with Possibilities of an Increase (−) or Decrease (--) in Incomes.

D. Income Profile for Free Lancers as Authors, Artists with fluctuating Higher and Low Incomes.

E. Income Profiles of Hazardous Occupational Personnel Having Higher Earning (−) but Short Period, but some Maintain to High Incomes with Accumulated Wealth (- - -).

There may be a difference in the use of income among two families of similar size, stage of family life cycle and other housing conditions. Their effective use of income will vary on account of -

- Age of the home maker i.e. young ones are less informed.
- Advertisements influencing young ones more than the older home maker.
- Knowledge and experience in handling family finances.

Time

Time availability is limited to 24 hours a day, for everyone, but the demand on that may be less, equal to or greater than that can be met, as is evident from figure 11.6:

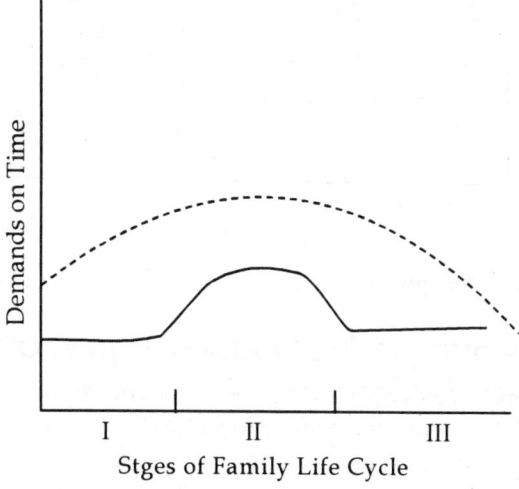

Fig. 11.6:   Time Demands During Family Life-cycle, Gainfully Employed Home-makers, (......) Non-gainfully Employed Home-maker (—)

For the beginning family i.e., in stage I, the demand on time is low for both the gainfully employed and the non-gainfully employed. The couple is busy in setting up the house and establishing the family. They may spend more time on household activities because either they are inexperienced or they have more time to set up the house. The families in stage II brings in a higher demands for time with the arrival of the first child. Care of the child takes away most of the time of the home-maker, besides trying to find time to shop for meeting the needs of the child. With the arrival of the second child, the time demands increase till the time the children grow up and become independent. In stage III, the mother finds more time for herself as the children leave their home . She might spend the extra time on taking up vocations or on doing social services. In joint families, she may be required to take care of her grand children, while the men may spend their time on leisure time activities like playing golf or bridge after their retirement from active employment pursuits.

Energy

Unlike the other sources, the flow of energy gives an interesting picture. Amount of energy flow and its demand follow in different courses. The supply of energy usually decreases as life advances. The demand upon it during stage I is light, heavy during Stage II and reduces once again in stage III, as shown in figure 11.7. This will once again vary among the gainfully employed and the full time home maker. Though the demand is light in stage I, it increases in stage II, where the supply of energy is found to be inadequate . Both frustration and physiological fatigue is common at this stage. In stage III, a great change occurs with the diminition of potential energy. This is likely to upset some women though the demands on energy may also come down.

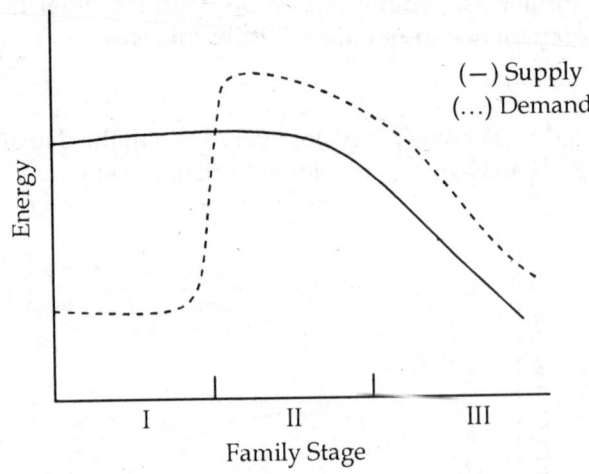

Fig. 11. 7: Possible Supply and Demands on Energy During Family Life Cycle

HANDLING DEMANDS DURING THE VARIOUS STAGES OF FAMILY LIFE CYCLE

Although certain problems in resource management are common to all stages of family life, the emphasis on each resource will vary from stage to stage in every family. Among these problems, the following requires thoughtful considerations:

- Recognising the importance of securing a suitable balance between work, rest and leisure time.
- Considering all members of the family in planning managerial activities
- Taking resource costs into consideration when making choices of goods and plans of action.
- Lowering resources costs to meet the extra demand in future.

An understanding of the above factors which is likely to affect resources uses during family life cycle,will help families to plan ahead and prepare themselves to meet new changing demands during the different stages.

Stage I: Beginning Family

This stage starts with the marriage of a man and a woman . This is the first sub stage of family life cycle, also known as the period of establishment. It begins with the marriage and continues until the first child is born. This stage is also a period of adjustment. It is also called the period of getting acquainted stage, when the two partners in the marriage are getting to know each other . The kind of relationships that the husband and wife establish at this time, will determine the quality of their future relationship. The long time goals they plan and realise at this time will influence their way of living and the manner in which they use their resources of time, energy and income throughout their married life.

It can thus be seen that in the first period of life cycle, there are not only managerial problems to be faced, but personal relationships also inter-mingle with managerial problems and decisions. During this period, long term goals, habits of work, division of responsibility and other living patterns are determined. While doing so, there may be possible conflicts over the use of resources, and therefore, the attitudes and feelings of the two partners towards these resources may need to be harmonised. When the partners manage to make mutual adjustments and share their responsibilities successfully, they are well prepared to take up additional responsibilities after the arrival of their children.

Methods of Dividing Responsibilities

An important early decision for young couples is to how responsibilities are shared, divided or delegated. Democrative decision making in the family, whether related to expenditure or to other family activities is perhaps more of an ideal than a reality. However, there will be many matters to be worked out in this period of establishment, although not all families can or will adopt the democrative method of decision making, it warrants consideration.

During this period, the major problem that most couples face is that of finance. Unlike the use of other family resources, many couples do come across a sort of disagreement as far as their finances are concerned inspite of their relatively adequate earnings. There is a definite evidence that after sufficient funds for minimum needs have been provided, attitudes towards money appear more important than the amount of money that determine family happiness. It is also seen that some married women at every income level feel that their incomes are insufficient as compared to married men who do not consider so. However, women with low incomes are better adjusted than those with middle or large incomes.

Establishing a Home

While establishing the family, the couple find that providing a home is a major problem. Though a number of wedding gifts are received, the young couple are forced to purchase a considerable amount of home furnishings, and major household equipments such as refrigerators, washing machines, microwave oven etc. However, their expenses on other items are comparatively low, since there are only two members to feed. Their clothing needs are already met for sometime while planning for their marriages. Expenses on recreation and home maintenance are comparatively low and therefore, there is scope for making savings and some investments which would enable them to meet the higher demands at a time in future when the children start arriving.

At this stage, it is important to decide how funds should be allocated among various categories of expenditure in the budget and among family members. The three possible methods of handling family funds among members are:

- Joint control
- Allowance, and
- Dole

There is probably a trend towards joint control of finances by husband and wife with some differences between socio-economic classes. It is usually observed that in the lower socio-economic group, the women dominate whereas in the middle class, the husbands dominate in the control of finances. Among higher income families, there appears to prevail some joint planning, though the expenses are generally controlled independently.

Both the allowance and dole systems are based on the dominance of single individual. When the dole system is followed, one of the family members controls the use of money resources. This person doles out or hands out when he feels that the requests of other family members are justified. While the entire burden of all decisions falls on this person, the other family members feel that they are subordinate to this person. It becomes difficult for the recipient to make any advance planning since he is uncertain as to when and how much money he is going to receive. Sometimes it might lead to unpleasant feelings between the two when the person asking for the money may not be aware of the financial situations. Sometimes by fulfilling all demands of the family members, the head of the family may not realise how genuine these demands are or may consider some of them as less important as compared to the other family members.

Allowance system may be considered as slightly better than the dole system, but it is still inferior to joint control system. It is similar in nature to the dole system in the sense that a single person has the control of money but periodically a specific amount or allowance is given to other family members for meeting certain expenses. For example, the husband might hand over a specific sum to his wife as allowance to meet certain household expenses . It then depends on the housewife to organise her expenses within the allotted sum or she might economise in certain goods to make a saving for herself. But when the prices of commodities are very high, or when the allowance is found to be insufficient to meet the expenses, the husband should be willing to consider this and suitably increase the allowance. Otherwise it might lead to unpleasantness between the husband and wife. Otherwise the wife should be willing to adjust her expenses and run the household on a lesser budget, if she appreciates that her husband may not be able to provide more allowance. However, this system has a definite advantage over the dole system as the housewife has an idea about the amount of money she is going to receive whereas there is some uncertainty in the case of dole system.

Furnishing the Home

Whichever system is followed, the family members should consider their basic needs first. These include facilities for eating, sleeping, proper lighting,sitting and storage. For a beginning family,initial investments on kitchen utensils, dining room furniture and wares, a comfortable bed and some storage facilities, specially a good wardrobe, would meet the basic needs. While doing so, the young family should consider the financial and aesthetic aspects of furnishing their homes. At the same time, they should also consider the care required and the future operating costs of their home in the long run i.e. whether they will be able to continue to do so even after the children are born and the family expands.

In making plans for furnishing the home, some important principles underlying their selection are to be considered. The young couple need to decide about the amount of money to be spent, which is usually one fourth of the house costs, the items to be purchased, the desired characteristics, and the approximate time of purchase. A well timed plan and a satisfactory blending of their different backgrounds and their expectations, are likely to help the couple to meet their basic and future demands with success. While doing so, vital resources like their abilities, skills, interests, time and energy would help them to make the home look attractive and functional even with the use of basic furniture and furnishings. The launching stage for the parental family coincides here with the young family in the period of establishment . The long term goals made now will influence their way of living and the manner in which they use their resources like time, energy, skills, and money throughout their married life.

Stage II: The Expanding Family

This stage covers a longer period than the previous one. Thus the expanding family stage includes a number of sub stages unlike the beginning stage, which has only one sub stage. This stage begins with the birth of the first child and ends when the last child leave the home. The first sub stage is designated as child bearing and pre school. Alongwith the goods that were acquired in the earlier stage, the family also accumulates more material goods during this stage. This in turn brings about the development of attitudes that influence future family living. During this period, the parents realise that it is necessary to make adjustments in their relationship with each other and their children. The changes in this sub stage follow at a rapid pace especially when the children are born at short intervals. The parents who face such situations, learn to quickly take on new and different responsibilities.

The next stage is the elementary school period, when the children are about 6-12 years of age. This brings in a series of overlapping experiences to the parents. It is at this time that children begin their formal education in the schools and make their first independent contacts with the outside world. During this period,the parents are primarily concerned with the educational and health needs of their children. They also get busy in establishing an environment in which the children will have a feeling of belonging, with providing nutritional foods, suitable clothing and adequate housing for personal development as well as social and community contacts outside the home. When children are included in the planning of family activities and budgeting, they show understanding and try to adjust their needs and demands accordingly.

The high school stage that comes after this stage, makes a demand on the parents to help the children to solve their personal problems like social, recreational, personality development etc. and to help them to become independent and self reliant . During the college phase, the children need parental guidance to take up a suitable area of subject specialisation in higher studies, and for financial arrangement or assistance to meet their college expenses. The parents may use some of their savings to meet such situations. The parents continue to help them to some extent, at the time of marriage and in establishing a new home for them.

The changes in the family patterns in the active parenthood stage of the cycle are due to

- An increased number of family members
- The growing feeling of responsibilities, and
- An increase in demand for most of the resources.

The changes may more or less remain the same because of the increase in the size of the family but may also vary correspondingly according to the age of the children . For example, when the children are small, they may demand more time and energy from their mother, but when they grow up and start studying in a college, the same children's demand would be more for money than for time. This demand will become more when the child lives away from the home and this sub stage becomes the most expensive one. Thus, in the expanding stage of family life cycle, the age of the child sets different patterns on the use of family resources.

Even though the family generally gets established by the beginning of this stage, most families find this stage more expensive than ever. In the child bearing and pre-school stage, health care and medical expenses are high . Besides. the mother also needs either a paid help or may need labour saving devices in her household, since the demand on her time increases on the arrival of the child. Housing costs may also go up if the family has a need for more space. Children also outgrow their dresses during the growing years and therefore there is always a continuous demand for expenses on clothing. Since children have tremendous amount of energy at this stage, some outdoor activities like sports, music or art pursuits may require additional expenses.

As the child grows, expenses on food also increases and the family may find this as an additional burden. When the child is in college stage, the family might have to withdraw from savings for meeting the extra money requirements. It is necessary for the married couple to realise that throughout these sub stages, there is a constant financial demand to meet the costs of health care, medical, child rearing and education.

Stage III: The Contracting Family

This period begins when the first child leaves the home. This stage is included under the sub-stage called the period of vocational adjustments and launching period. The child may leave the home either to join a service or to take up a career or to get married. The period of financial recovery sets in when all the children leave home and become self supporting. The family expenses drop and savings may accumulate or the old debts, if any, are settled now. Once this is done, any saving the family does is for the retirement or for the old age period.

The next stage i.e., the period of financial recovery, calls for the social and vocational adjustments. Since children are not there, the husband and wife have to adjust their time in a manner that they can spend more time on neighbourhood, social or community activities instead of family activities. This would provide them an opportunity to adjust themselves in the absence of children in the home.

As mentioned earlier, this stage of the family life-cycle has three sub- stages. The first sub-stage is the launching period during which the children are leaving the home, the second is the recovery period, when the children starts earning and leave the home, and finally the ageing or the retirement period when he is no longer employed. Each of these three sub-stages has different and distinct characteristics.

The launching period, also known as the period of vocational adjustment of the children may come immediately after high school stages in some cases whereas it may start only after college stage with others. This is the time when children start settling in their professions. Parents may offer either guidance or financial assistance or both. In the case of a family having a daughter might call for additional expenses in getting her married and then settling her in the newly found home. Every family is aware of the additional costs of the marriage and the demand on time and energy resources involved in celebrating the marriage of a daughter. However, every member in the family willingly either contribute their share or do the best of their capacity and see that the daughter is well settled in her new home.

When the children are established either in their profession or marriage, as the case may be, the parents enter the seventh sub stage, called the stage of financial recovery. If the family did not have enough savings, it might have accumulated some debts. Such families need to focus on returning their borrowings, besides preparing savings for their retirement period, when money may not flow in, or only a limited or meagre amount as pension may be received. Thus the third major stage of family life cycle assumes greater importance on account of these two demands. They may have more time for themselves, but their energy is reduced because of their age.

During the retirement period, the greatest problem every family faces is that of reduced income or no income. A family therefore, need to make enough savings to meet the requirements of this stage. This is more important now more than ever since the concept of nuclear family over the joint family is best catching up in the Indian society. The parents, therefore, do not get the support of their children either financially or physically or both in some cases, though the demands for these resources may not be much. The only extra expenditure might be that of medical and health care during this stage. Availability of time resources might be there in plenty, but the older couple need to focus on engaging themselves in some occupations or sports, or social services. Some may offer their services for religious activities.

Thus this stage calls for a lot of adjustment on the part of the older couple. Some may also accept a full time job at a lesser renumeration or part time jobs, which again calls for some minor adjustments. Some may also engage themselves in creative activities like writing books, or in recreational activities like gardening which might be more satisfying to them than accepting a salaried job. Some of them may take up such assignments or hobbies or interests which they could not take up in their earlier days. However, the final decision on such options will depend upon their abilities, health status, income and the space available.

Most people find living in their own homes, more enjoyable than to live in someone else's house which might give them a feeling of being dependant, or make them feel as if they would loose their prestige or authority. This might be felt intensively specially when living space is limited. Since such situations are likely to create conflicts between children or grand children and the grand parents, there appears to be a need for more old age homes than ever before. It is generally felt that women are able to adjust better than men in these circumstances. The problem becomes more acute when only one member of the couple is alive and is forced to depend on the children for support, who were themselves dependant on him in earlier days.

Thus, it can be seen that in the last sub stage of retirement, problems like diminished income, failing health, reduced energy and extra time, have to be tackled with patience. If they have saved enough money in the past, they would not require any monetary support from their children. Some part- time jobs or social service or leisure time activities would keep them occupied. Even a small income may be acceptable, as this will keep them occupied.

IMPORTANCE OF UNDERSTANDING FAMILY LIFE CYCLE

The couple, at the time of marriage, are not in a position to foresee and plan for a secure and trouble free future. Therefore, knowledge of the life cycle of the family during the different stages, helps a family in planning for the uncertainties in the future. It enables them to look ahead and anticipate the needs and demands during the subsequent stages. The family is well prepared and is not taken unaware when it can foresee and understand the possible happenings during the various stages of family life cycle.

Bibliography «««←

Abbass, F. and Alkhafaji. (1992). *Competitive Global Management – Principles and Strategies*. New Delhi: Vanity Book International.

Adaviyappa, R. (1976). *Factors Affecting the Utilisation of Time in Various Household Activities by Urban Home Makers*. Unpublished master's thesis submitted to M. S. University, filed at Hansa Mehta University Library, Boroda.

Agan, T. (1967). *The House: Its Plan and Use*. New Jersey; Prentice Hall Inc.

Agnihotry, P. (1996). *Value Orientation of Doctors: A Comparative Study Amongst Private and Public Hospitals in Delhi*. Unpublished dissertation, Faculty of Management Studies, New Delhi.

Alhandnis, U. and Malini, I. (1987). *Women of the World – Illusions and Reality*. New Delhi; Vikas Publishers.

Allen, S. K. and Wolkowitz, C. (1987). *Homeworking: Myths and Realities*. London; Macmillan.

Allport, G., Vernon, P. and Lindzey, G. (1960). *A Study of Values*. Boston; Houghton- Miffin.

Amarchand, A. and Jayaraj, B. J. (1992). *Corporate Culture and Organisational Effectiveness*. New Delhi; Global Business Press.

Amba-Rao Sita, C. (1979, Nov.). The managerial mainstream and Indian women. *Indian Management*, **18**.

Aplander, G. Guvence and Gutmann, E. Jean. (1976, Feb.). Contents and techniques of management development programme for women. *Personnel Journal*, **55**.

Batra, S. and Seetharaman, P. (2000). *Perception and Utilization Pattern of Time among Urban Women in Different Occupations*. Unpublished doctoral thesis of Delhi University. Delhi.

Bergson, H. (1910). *Time and Free Will*. Ph.D., Pogsor London: George, Allen and Unwin.

Bruke and Goodstein (1980). *Trends and Issues in Organisational Development: Current Theories and Practice*. California; University Associates Inc.

Casay T. W. (1979). Asian values in management. *Indian Management*, **18**(5).

Chakraborty, S. K. (1991). *Management by Values*. New Delhi; Oxford University Press.

Chakraborty, S. K. (1993). *Management by Values: Towards Cultural Congruences*. Delhi; Oxford University Press.

Chakraborty, S. K. (ed.). (1995). *Human Values for Managers*. New Delhi; A. H. Wheeler and Co. Ltd.

Chandra, A. (1995). *Introduction to Home Science* (2nd ed.). New Delhi; Metropolitan Book Co. Pvt. Ltd.

Chandra, D. (1977). *A Study of Perception of Work Values in Teaching and Certain Non-Teaching Occupations*. Unpublished doctoral thesis, Aligarh Muslim University, Aligarh.

Chattergee, S. (1982). *An Introduction to Management – Its' Principles and Techniques*. Calcutta; The World Press Private Limited.

Chowdhury, B. K. R. (1996). Motivating factors in workers: A study relating difference in the opinions of male and female managers. *Indian Management*. September.

Conner, P. E. and Becker, B. W. (1975). Values and organisation: suggestions for research. *Academy of Management Journal*, **18**: 3.

Cowles, L. and Deitz, R. P. (1956). Time spent in homemaking activities by a selected group of Wisconsin Farm Homemakers. *Journal of Home Management.* **48**(1).

Daftuar, C. N. (1982). *Job Attitudes in Indian Management.* Delhi; Naurang Rai Concepts Publication Company.

Deacon, E. C. and Bratton, P. E. (1954). Home management focus and function. *Journal of Home Economics.*

Dessai, N. and Anantram, S. (1985). Middle class women's entry into the world of work. *In*: K. Saradomoni (ed.). *Women, Work and Society.* Proceedings of ISI Symposium, Indian Standard Institute, Calcutta.

Dhingra, S. (1984). *A Comparative Study of Time-use Pattern of Women Engaged in Home Based and Factory Based Work.* Unpublished master's thesis, Delhi University.

Drucker, P. (1977). *People and Performance.* New Delhi; Allied Publishers Pvt. Ltd. Employment and Unemployment in India. (2000). NSSO Report.

England, G. W. (1974). Relationship between managerial values and managerial success in U.S., Japan, India and Australia. *Journal of Applied Psychology,* **59**(4).

Fleck, H. (1968). *Towards Better Teaching of Home Economics.* New York; Macmillan Co. Ltd.

Fogerty, M. (1971). *Women in Top Jobs.* George Allen and Unwin Ltd. Ruskin House.

Fraisse, P. (1963). *The Psychology of Time.* New York; Harper and Row.

Gautam, V. (1992). *Issues in Organisational Management: Prediction and Potential.* New Delhi; Global Business Press.

Gollapudi, S. R. (1991). *Contemporary Studies in Organisational Behaviour.* V. S. P. Rao (ed.). New Delhi; Discovery Publishing House.

Gore, M. S. (1968). *Urbanization and Family Change.* Bombay; Popular Prakasana.

Gradly, E. R. (1962). Management and Mental Health. *Journal of Home Economics.*

Graves, C. W. (1970). Levels of existence: An open system theory of values. *Journal of Humanistic Psychology,* no. 2.

Gross, I. H. and Crandall, E. W. (1954). *Management for Modern Families.* Delhi; Sterling Publishers Pvt Ltd.

Gross, I. H. Crandall, E. W. and Knoll, M. M. (1973). *Management for Modern Families.* New Jersey; Prentice Hall Inc.

Guaba, A. (1985). *Wife-Husband Participation at Home.* Paper presented at the Asian Reginal Conference on Women and the Household, New Delhi.

...ti, M. (1982). *A Study on Use of Time by Non-Working Homemakers and their Attitude Towards Time Management.* An abstract of the unpublished master's thesis, M.S. University, Baroda.

...ta, N. K. and Ahmad, A. (1994). *Managing Transition.* New Delhi; New Age International Publications.

...ode, S. (1964). *Impact of the Arrival of Children on Time and Money Management of Fifty Homemakers in Coimbatore City.* Unpublished master's thesis. Research in India related to Home Management, a compilation of research findings by Walls, H.L.

...ndel, H. M. (1951). Time estimates as a function of distance travelled and relative clarity to goals. *Journal of Psychology.*

...lindle, T. (1998). *Manage your Time.* London; Dorling Kindersley Limited.

...lofstede, G. (1980). *Culture's Consequences.* Beverly Hills; Sage Publications.

Hofstede, G. (1991). *Cultures and Organisations: Software of the Mind.* Berkshire, UK; McGraw Hill.

Indian Census Report (1991). Government of India.

Ingersoll, H. L. (1961). Family values in consumer age. *Journal of Home Economics.* September, **53**(7).

Jain, D. and Chand, M. (1982). *Report on a Time Allocation study, its Methodological Implications.* Paper presented a the technical Seminar on Women's Work and Employment, New Delhi, ISST.

Janu Z, L. R. and Jones, S. K. (1982). *Time Management for Executives.* Britain; Sidgwick and Jacksons Ltd.

Kakkar, S. (1979). *Indian Childhood.* New Delhi; Oxford University Press.

Kala Rani. (1976). *Role Conflict in Working Women.* New Delhi; Chetna Publications.

Kapoor, P. (1974). *The Changing Status of Working Women in India.* Delhi; Vikas Publications.

Koontz, O'Donnell, Weihrich. (1986). *Essentials of Management.* (4th ed.). New Delhi; McGraw Hill Book Co.

Kraisanawasdi, N. (1989). *Women Executive: A Sociological Study in Role Effectiveness.* Jaipur; Rawat Publication:

Kumari, P. (1963). *Contribution of Various Family Members Towards Home Making Activities in Selected Househol in Madras City.* Unpublished doctoral thesis, University of Madras, Madras.

Lazlo, E. (1973). A Systems Philosophy of Human Values. *Behavioural Science,* no. 18.

Leonard, W. E. and Thakkar, U. (1989, June). The management development of women administrations: some considerations. *University News*, **27**.

Maggu, P. , Seetharaman, P. and Khanna, K. (2000). *Urban Kitchens and Meal Preparation Activities-An Assessment of their Association*. Unpublished doctoral thesis of Delhi University. Delhi.

Magrabi, F., Paolucci, R. K. and Heifner, M. E. (1967, Nov.). Framework for studying family activity patterns. *Journal of Home Economics*, **59**(9).

Masini, E. and Strattibas, S. (1991). *Women, Household and Change*. Japan; United National University.

Mehra, P. and Seetharaman, P. (2003). *Management Training and Values: A Study on Women Managing Formal and Informal Organisations Simultaneous*. Unpublished doctoral thesis of Delhi University. Delhi.

Miller, P. L. (1961). Home management patterns of three generations. *Journal of Home Economics*. February, **53**(2).

Nickell, P. and Dorsey, J. M. (1991). *Management in Family Living*. New Delhi; Winley Eastern Ltd.

Nickell, P., Rice, A. and Tucker, S. (1976 & 1978). *Management in Family Living*. New York; John Wiley Sons.

Nickols, S. Y. and Nickols, S. A. (1983). *Families of the Future: Continuity and Change*. College of Home Economics, Ames; IOWA State University Press.

O'Connell, H. (1994). *Women and the Family*. London; Zed Books.

Parnell, F. B. (1981). *Homemaking Skills for Everyday Living*. South Holland, Illinois; The Good Heart-Willoy Company, Inc.

Parsons, T. (1964). *Social Structure and Personality*. New York; Free Press.

Powell, G. N. (1988). *Women and Men in Management*. New Delhi; Sage Publications.

Prasad, L. M. (1994). *Organisational Behaviour*. New Delhi; Sultan Chand and Sons.

Rai, C. M. (1977). Value System of Indian Managers. *Indian Management*. February.

Rama Rao, P. (1971). *Study of Relationship Between Certain Factors of Personality and Time Perception*. Unpublished doctoral thesis, Sri Venkateshwara University.

Rao, P. R. (1978). *Studies in Time Perception*. Delhi; Concept Publishers.

Robbins, S. P. (1996). *Organisational Behaviour*. (7th ed.). New Delhi; Prentice Hall of India Pvt. Ltd.

Rokeach, M. (1968). *Beliefs, Attitudes and Values*. San Francisco; Jossey Bass.

Rokeach, M. (1973). *The Nature of Human Values*. New York; The Free Press.

Saikia, R. (1979). *A Study on Managemenet Practics of the Homemakers of Jorhat as Reflected in the Use of Time and Money*. Abstract of unpublished master's thesis, M.S. University, Baroda.

Sarkar, A. N. (1985). The Need for an Indianised Management System. *The Management Review* III, **12**(3).

Schein, E. H. (1986). *Organisational Culture and Leadership*. London; Jossey Bass Publishers.

Seetharaman, P. and Sethi, M. (2002). *Consumerism Strategies and Tactics*. New Delhi; CBS Publishers and Distributors.

Selvaraj, M. U. (1998). Women in Management. *Third Concept*. December.

Seetharaman, P. and Pannu, P. (2004). *An Introduction to Interior Decoration*. New Delhi; CBS Publishers and Distributors.

Sethi, N. K. (1992). Changing values and considerations in India. *In: Managerial Dynamics*. New Delhi; Sterling Publication Pvt. Ltd.

Sethi, M. (2004). *Institutional Food Management*. New Delhi; New Age International Publishers.

Singh, P. (1979). *Occupational Values and Styles of Indian Managers*. New Delhi; Wiely Eastern.

Singh, S. K. (1990). Value Preference and Choice of Ingroups. In *Organisational Research in Indian Perspective*. New Delhi: Northern Book Center.

Sinha, D. and Rao, S. R. (ed.). (1988). *Social Values and Development: Asian Perspectives*. Newburypark, California: Sage publications.

Sinha, D. P. (1995-96). Culture and Management. *Organisational Management*, **xi** (3-4).

Sinha, J. B. P. (1987). *Work Culture in Indian Work Organisation*. ICSSR Report.

Singh, Nirmal. (2000). *Principles of Management – Theory, Practices Techniques*. New Delhi; Deep and Deep Publishers.

Soares, F., Valecha, G. K. and Venkataraman, S. (1992). Values of Indian managers: The basis for progress. *Productivity*, July.

Steidle, R. G. and Bratton, E. C. (1968). *Work in the Home*. New York; John Wiley.

Stewart, N. (1978). *The Effective Women Manager*. New York; John Wiley & Sons.

Super, D. E. (1970). *Work Value Inventory*. Chicago; Riverside.

Swanson, B. (1981). *Introduction to Home Management*. New York; Macmillan.

The 25 Most Powerful Women in Indian Business. (2003, Nov. 23). *Business Today*.

Thiagarajan, K. M. (1974). Mutual perception of manager's and worker's values – A cross cultural study. *Indian Management*. December.

Unwalla, J. M. (1977). *Beyond the Household Walls – A Study of Women Executives at Work and at Home*. Unpublished doctoral thesis, Bombay University.

Upadhyay, D. P. (1985). Value, people and organisation. *Indian Management*. vol, December.

Uris, A. and Noppel, M. (1970). *The Turned-on Executive*. U.S.A.; McGraw-Hill Inc.

Verma, N. (1995). *Exploration in Time Use*. New Delhi; Classical.

Warrier, S. K. (1983, July). Value profile of Indian managers. *Indian Management*.

Yankellovich, D. (1994). How changes in the economy are reshaping American values. *In: Values and Public Policy*. Washington; The Brooking Institute.

Index ◀◀◀◀—

4-point scale, 75
Abbass, F., 4
Acceptance needs, 68
Activity plans, 133, 136
    steps in making, 136-37
Alkhafaji, 4
Amrita Patel, 20
Assured income, 182
Aversion, 158
Batra, S., 127-28
Bending, 155
Betty Stevenson, 176
Betty. B. Swanson, 103
Biological time, 125
Body mechanics, 150
Bonde, 85
Borae O Sale Birg, 19
Boredom, 156, 157
Boredom fatigue, 157-58
    elimination of, 159
Brech, 11, 12
Budgetary control, 59
Budgeting, 17-18, 179
    limitations of, 179-80
    necessity of, 179
    steps in making, 180-84
Business management, 13
Cash budget, 59
Centre of gravity, 153-54
Centre of weight, 153-54
Chakraborthy, 72
Chanda Kochhar, 20

Chang, 216
Chitta-Siddhi, 72
Chowk Paylay, 216
Clock time, 125
Conflict, 98-99
    methods of resolving, 99-100
Contemporary Indian values, 73-74
Contingency theory, 37
Controlling, 57
    characteristics of, 57
    process of, 57-58
    role of, 58-59
    techniques of, 59
Core values, 72
Cornell, 129
Cowels, 127
Crandall, E. W., 11, 44, 51, 70, 89, 95, 103, 105, 162, 174,
Credit, 192
    sources of, 193
    types of, 192
    use of, 194
Culture, comparative values of, 6
    differences of, 6
    dimensions of, 4- 5
Deacon, 44, 103
Decision making, 9, 88
    aids in, 101-102
    complexities of, 100
    definitions of, 89
    factors affecting, 95
    integration of, 91

process of, 9
relation of, 89
role of, 90
steps of, 92-95
Decision tree, 98
Decision, kinds of, 96-98
Decisional role, 25
Deliberate decision, 97
Derivative plan, 47
Devis, 50
Directing, 56
Disseminator role, 25
Dissonance, 95
Dorsey, J. M., 11, 173, 221
Drucker, P., 23
Dry rangoli, 218
Elton Mayo, 36
Emergency decisions, 98
Energy plan, control of, 169
Energy resource, 146
Energy spending patterns, 149
England, G. W., 75
Esprit De. Corps, 43
Esteem needs, 68
Esther Crew Bratton, 151
Evaluation, 63
    characteristics of, 63
    purposes of, 63
Everyday activities, 136, 138, 140
Expenditure budget, 59
Family development, 8
    objective of, 8
Family income, 173-74
    classification of, 174, 176
    control of, 184-85
    record keeping, 185
    sources of, 174
    supplementing, 177-78
    types of, 174
Family life cycle, 142, 146, 221
    analysis of, 229
    concept of, 227
    environment surrounding, 222
    handling demands, 236
    importance of, 228, 241
    management of resources in, 233
    recent trends in, 230, 232
        impact of, 232
    stages of, 142, 146-47, 228
Family living, 7

management of, 7
    role of homemaker in, 7
Family, 221, 223
    concept of, 223
    functions of, 224-225
    joint, 223
    matriarchal, 224
    nuclear, 223
    patriarchal, 224
    role of, 224
Fatigue, 155
    causes of, 155
    remedies of, 155
    types of, 155-56
Fayal, H., 12, 36
    contributions of, 40
    management principles of, 40-42
Figure head, 24
Firebaugh, 103
Floral rangoli, 219
Flower arrangement, 211
    leading styles, 213-14
    steps of, 212-13
Follet, 16
Frederick W. Talyor, 36
    contributions of, 39
    management principles of, 39
Frustration, 156-58
Frustration fatigue, 158
    reasons of, 158
    remedies of, 160
    elimination of, 159
Functional space, 201
Functional storage, 206
    guidelines of, 206
Gender bias, 21
Goals, 79, 86-87
    role of, 80- 81
Greenberg, 156
Gross, I. H., 11, 44, 51, 70, 89, 95, 103, 105, 162, 174
Group decision, 96-97
Guatum, 72
Gudjansson, 12
Guiding, 56
Haqqard, 156
Hazelton, 155,
Henry M. Boettinger, 17
Herzberg, 68
Hidden income, 176
Hill, 224

Hofstede, 4-6, 71
  concept of culture, 4
  dimension of, 4
Home accessories, 210
  arrangement of, 210-11
  placement of, 211
  selection of, 210
Home establishment, 237
Home management, 8-9, 10-11, 13
  characteristics of, 8-9
  definitions of, 11-12
  ethics in, 21-22
  misconceptions of, 15-16
  purview of, 9
  quality of, 8-9
  role of, 9-10
  system approach to, 37
Home management skill, 8-9
  transfer of, 9
Homemaker, 9, 11
  activities of, 7
  characteristics of, 29-30
  decisional roles of, 27
  functions of, 30-31
  informational roles of, 26
  interpersonal roles of, 26
  qualities of, 27-29
  responsibilities of, 26
  role of, 26
Homemaking activities, 147
  efforts used in, 147-48
Hort, L. C., 11
House furnishing, 200, 238
  arranging, 200
  choosing, 200
Household activities, 149
  classification of, 149-50
Household environment, 222
Household pests, 207
  control of, 208
Housing norms, 199
Human resources, 112
Ikebana, 214
Income budget, 59
Income, 173
  concept of, 173
  definition of, 173
  real, 173
Indian Census, 20
  report of, 20

Indian values, 72-73
Individual decision, 96-97
Industrial revolution, 6-7
Informational role, 25-26
Instrumental values, 70
Intermediate goals, 79-80
Interpersonal roles, 24-26
Intrinsic values, 70
Investment, 190
  avenues of, 191
  criteria of, 190, 191
Jet lag, 125
Kalash, 216
Kalpana Morparia, 20
Kathryn Walker, 129
Khanna, K.. 165
Khera, 216
Killion, R. A., 89
Kimball, 50
Kiran Mazumdar Shaw, 20
Kishori. J. Udeshi, 20
Knoll, 103, 162
Knowles, 158
Koontz O'Donell, 17
Kotzin, 12
Kyrk, 85
Lalita Gupte, 20
Le Briton, 44
Leading, 54-55
  characteristics of, 55
  roles of, 55-56
Leisure time, 128
  norms of, 128
Liaison, 24
Living standard, 84-85, 87
Longdon, 157
Long-term goals, 80
Lutter Gullick, 17
Magu, P., 165
Mahatama Gandhi, 118
Maloch, 44
Management process, activities of, 43, 45
Management, 1-2
  activities of, 89
  characteristics of, 18
  concept of, 7
  definition of, 1,
  general principles of, 19
  growth of, 35
  history of, 35

home, 1
  importance of, 23
  need of, 6-7
  philosophy of, 1-2
  principles of, 35, 39
  process as applied in home, 8
  process of, 35
  process of, 43
  roll of, 23
  schools of, 36-37
  universality of, 2-3
Manager, 23-24
  functions of, 30-31
  leadership role of, 25
  performance of, 23
    measurement of, 23
  role of, 24-25
  style of, 33
  types of, 32-33
Maslow, A. H., 68
Massic, J. L., 89
Mayo, 159
Mean end goals, 80
Mehra, P., 77
Meloch, 103
Mental effort, 147-148
Money income, 174, 176
  sources of, 174
Money management, 173
  steps of, 178
Monitor role, 25
Morris Cooke, 36
Motion mindedness, 161
Motivation, 66
  factors of, 69
Motivator-Hygience approach, 68
Naina Lal Kidwai, 20
Near environment, 222
Needs theory, 68
  hierarchy of, 68
Need-want-satisfaction chain, 67
Net Worth statement, 63
Nickell, P., 11, 173, 221
Nirmal Singh, 44, 51
Non-budgetary control, 59
Non-human resources, 112-14
Normative values, 70
Oliver Sheldon, 50
Operation chart, 161, 164
  symbol of, 164

Operative plan, 47
Organisation, 4, 12, 23
  complexity of, 12-13
  cultural characteristics of, 4
  cultural value of, 3-4, 6
  formal, 3
  goals of, 13
  informal, 3
  motive of, 12
  semiformal, 3
  size of, 12
  type of, 12
Organised set of behaviour, 24
Organising, 50-54
  characteristics of, 51-52
  importance of, 53-54
  process of, 52-53
Pannu, P., 210
Parker, D. H., 70
Path process chart, 165
  symbol of, 166-68
Pathway chart, 161-62
Payne, R., 3
  concept of culture, 3
Peak loads, 130, 135
Pedal effort, 147-48
Periphery values, 72
Physiological fatigue, 156
  alleviation of, 159
Physiological needs, 68
Planning, 44
  characteristics of, 44-45
  definition of, 44
  key elements of, 44
  limitations of, 49-50
  process of, 45-48
  role of, 48-49
Planning premises, 46
  development of, 46
Possible income, 182
Posture, 154-55
Process chart, 161-63
  symbol of, 162, 164
Productive income, 176
Professionalism, 18-19
  characteristics of, 19
Profit and lose account, 63
Psychic income, 175, 177
Psychological fatigue, alleviation of, 159
Psychological time, 125

Pulling, 155
Pushing, 155
Rangoli, 215
    patterns of, 216-218
    techniques of, 218-19
Rank order form, 77, 127
Real income, 174
    types of, 175
Relaxation, 159
Renuka Ramanath, 20
Resources, 103
    characteristics of, 104-07
    classification of, 109-11
    conservation of, 120
    definitions of, 103-04
    maximizing the use of, 119-20
    meaning of, 103
    role of, 104
    use of, 108
    use of, 116
        factors affecting, 116-18
Resource cycle, 229
Rest periods, 131, 159
    frequency of, 132
    length of, 132
Rokeach, M., 71
Russell, 155
Ryan, 158
Rythm, 155
Safety needs, 68
Sanitation, 209
Sarparveen Kaur, 20
Savings, 188
    advantages of, 189-90
Schien, E. H. 18-19
Scientific management, 36
    father of, 39
Seasonal activities, 137-39
Seetharaman P., 77, 127-28, 164, 194, 210
Shared resources, 112, 114, 115
Shikha Sharma, 20
Singh, P., 71, 89
Small family, 232
Solah Deepak, 216
Sorokin, 129
Space management, 195
    evaluation of, 219
Space planning, 195-96
    steps of, 197
Spriengel, 51

Staffing, 54
    activities of, 54
Standard, 82, 86-87
    change in, 86
    classification of, 82
    factors affecting, 84-85
Storage, 204-05
    advantages of, 204-05
    principles of, 205-06
Strategic planning, 48
Stress, 171
Stress related fatigue, 171
Subjective fatigue, 156
Sumita Narain, 20
Supervising, 56
    types of, 56
System theory, 37
Time, 124
    management of, 124
    management, importance of, 124
    nature of, 124
    norms of, 127-28
    use of, 125
Time demands, 142
Time patterns, 126, 134
Time plan, 133, 144
    bases of, 133
    control of, 143
    evaluation of, 144
    factors to consider in making, 135
    value of, 133
Time schedules, 138
    preparation of, 140-41
Tiring activities, 149
Tkonoma, 215
Torsal effort, 147-48
Unbiased decision, 21
Urwick, 12
Value scales, 75-76
Value systems, 75
Values, 69-70, 86-87
    changes in, 77-78
    classification of, 70
    dimensions of, 71
    significance of, 77
Vineeta Rai, 20
Visual effort, 147-48
Warming up, 131-32
Weekly activities, 136, 138-40

Weigand, 127-29
Weighted mean scores (WMS), 77
Wet rangoli, 218
Wielrich, 17
William F. Glueck, 50
Work curve, 130, 132, 157
Work participation rate, 20

Work space, 201
   designing of, 201
Work units, 129-30
   definition of, 129
Working area, 202-03
Workplace address, 166
Wyatt, 157